ALBERT R. MANN
LIBRARY
AT
CORNELL UNIVERSITY

ANCIENT PLANTS

Photo. of the specimen in Manchester Museum.

THE STUMP OF A *LEPIDODENDRON* FROM THE COAL MEASURES

ANCIENT PLANTS

BEING A SIMPLE ACCOUNT OF THE
PAST VEGETATION OF THE EARTH
AND OF THE RECENT IMPORTANT
DISCOVERIES MADE IN THIS REALM
OF NATURE STUDY

BY

MARIE C. STOPES, D.Sc., Ph.D., F.L.S.

Lecturer in Fossil Botany, Manchester University
Author of "The Study of Plant Life for Young People"

LONDON

BLACKIE & SON, Limited, 50 OLD BAILEY, E.C.
GLASGOW AND BOMBAY
1910

Preface

The number and the importance of the discoveries which have been made in the course of the last five or six years in the realm of Fossil Botany have largely altered the aspect of the subject and greatly widened its horizon. Until comparatively recent times the rather narrow outlook and the technical difficulties of the study made it one which could only be appreciated by specialists. This has been gradually changed, owing to the detailed anatomical work which it was found possible to do on the carboniferous plants, and which proved to be of great botanical importance. About ten years ago textbooks in English were written, and the subject was included in the work of the honours students of Botany at the Universities. To-day the important bearing of the results of this branch of Science on several others, as well as its intrinsic value, is so much greater, that anyone who is at all acquainted with general science, and more particularly with Botany and Geology, must find much to interest him in it.

There is no book in the English language which places this really attractive subject before the non-specialist, and to do so is the aim of the present volume. The two excellent English books which we possess, viz. Seward's *Fossil Plants* (of which the first volume only has appeared, and that ten years ago) and Scott's *Studies in Fossil Botany*, are ideal for advanced University students. But they are written for students who are supposed to have a previous knowledge of technical botany, and prove very hard or impossible reading for those who are merely acquainted with Science in a general way, or for less advanced students.

The inclusion of fossil types in the South Kensington syllabus for Botany indicates the increasing importance attached to palæobotany, and as vital facts about several of those types are not to be found in a simply written book, the students preparing for the examination must find some difficulty in getting their information. Furthermore, Scott's book, the only up-to-date one, does not give a complete survey of the subject, but just selects the more important families to describe in detail.

Hence the present book was attempted for the double purpose of presenting the most interesting discoveries and general conclusions of recent years, and bringing together the subject as a whole.

The mass of information which has been collected about fossil plants is now enormous, and the greatest difficulty in writing this little book has been the necessity of eliminating much that is of great interest. The author awaits with fear and trembling the criticisms of specialists, who will probably find that many things considered by them as particularly interesting or essential have been left out. It is hoped that they will bear in mind the scope and aim of the book. I try to present only the structure raised on the foundation of the accumulated details of specialists' work, and not to demonstrate brick by brick the exposed foundation.

Though the book is not written specially for them, it is probable that University students may find it useful as a general survey of the whole subject, for there is much in it that can only be learned otherwise by reference to innumerable original monographs.

In writing this book all possible sources of information have been consulted, and though Scott's *Studies* [1] naturally formed the foundation of some of the chapters on Pteridophytes, the authorities for all the general part and the recent discoveries are the numerous memoirs published by many different learned societies here and abroad.

As these pages are primarily for the use of those who have no very technical preliminary training, the simplest language possible which is consistent with a concise style has always been adopted. The necessary technical terms are either explained in the context or in the glossary at the end of the book. The list of the more important authorities makes no pretence of including all the references that might be consulted with advantage, but merely indicates the more important volumes and papers which anyone should read who wishes to follow up the subject.

All the illustrations are made for the book itself, and I am much obliged to Mr. D. M. S. Watson, B.Sc., for the microphotos of plant anatomy which adorn its pages. The figures and diagram are my own work.

This book is dedicated to college students, to the senior pupils of good schools where the subject is beginning to find a place in the higher courses of Botany, but especially to all those who take an interest in plant evolution because it forms a thread in the web of life whose design they wish to trace.

M. C. STOPES.

December, 1909.

---

[1] My book was entirely written before the second edition of Scott's *Studies* appeared, which, had it been available, would have tempted me to escape some of the labour several of the chapters of this little book involved.

Contents

Chap.		Page
I.	Introductory	1
II.	Various Kinds of Fossil Plants	6
III.	Coal, the most Important of Plant Remains	22
IV.	The Seven Ages of Plant Life	33
V.	Stages in Plant Evolution	43
VI.	Minute Structure of Fossil Plants — Likenesses to Living Ones	53
VII.	Minute Structure of Fossil Plants — Differences from Living Ones	69
VIII.	Past Histories of Plant Families— (i) Flowering Plants	79
IX.	,, ,, (ii) Higher Gymnosperms	86
X.	,, ,, (iii) Bennettitales	102
XI.	,, ,, (iv) The Cycads	109
XII.	,, ,, (v) Pteridosperms	114
XIII.	,, ,, (vi) The Ferns	124
XIV.	,, ,, (vii) The Lycopods	133
XV.	,, ,, (viii) The Horsetails	145
XVI.	,, ,, (ix) Sphenophyllales	153
XVII.	,, ,, (x) The Lower Plants	161
XVIII.	Fossil Plants as Records of Ancient Countries	168
XIX.	Conclusion	174

CONTENTS

APPENDIX

		Page
I. LIST OF REQUIREMENTS FOR A COLLECTING EXPEDITION	-	183
II. TREATMENT OF SPECIMENS	-	184
III. LITERATURE - -		186
GLOSSARY - - -		188
INDEX - - -		193

ANCIENT PLANTS

CHAPTER I

INTRODUCTORY

The lore of the plants which have successively clothed this ancient earth during the thousands of centuries before men appeared is generally ignored or tossed on one side with a contemptuous comment on the dullness and "dryness" of fossil botany.

It is true that all that remains of the once luxuriant vegetation are fragments preserved in stone, fragments which often show little of beauty or value to the untrained eye; but nevertheless these fragments can tell a story of great interest when once we have the clue to their meaning.

The plants which lived when the world was young were not the same as those which live to-day, yet they filled much the same place in the economy of nature, and were as vitally important to the animals then depending on them as are the plants which are now indispensable to man. To-day the life of the modern plants interests many people, and even philosophers have examined the structure of their bodies and have pondered over the great unanswered questions of the cause and the course of their evolution. But all the plants which are now alive are the descendants of those which lived a few years ago, and those again came down through generation after generation from the plants which inhabited the world before the races of men existed. If, therefore, we wish to know and understand the vegetation living to-day we must look into the past histories of the families of plants, and there is no way to do this at once so simple and so direct (in theory) as to examine the remains of the plants which actually lived in that past. Yet when we come to do this practically we encounter many difficulties, which have discouraged all but enthusiasts from attempting the study hitherto, but which in reality need not dismay us.

When Lindley and Hutton, in 1831, began to publish their classical book *The Fossil Flora of Great Britain*, they could give but isolated fragments of information concerning the fossils they described, and the results of their work threw but little light on the theoretical problems of morphology and classification of living plants. Since then great advance has been made, and now the sum of our knowledge of the subject, though far from complete, is so considerable and has such a far-reaching influence that it is becoming the chief inspiration of several branches of modern botany. Of the many workers who have contributed to this stock of knowledge the foremost, as he was the pioneer in the investigations on modern lines, is Williamson, who was a professor at Manchester University, and whose monographs and specimens are classics to-day. Still living is Dr. Scott, whose greatness is scarcely less, as well as an ever-increasing number of specialists in this country, who are continually making discoveries. Abroad, the chief Continental names are Renault, Bertrand, Count Solms Laubach, Brongniart, Zeiller; and in America is Dr. Wieland; while there are innumerable other workers in the field who have deepened and widened the channels of information. The literature on fossil plants is now vast; so great that to give merely the names of the publications would fill a very large volume.

But, like the records left by the plants themselves, most of this literature is unreadable by any but specialists, and its really vital interest is enclosed in a petrifying medium of technicalities. It is to give their results in a more accessible form that the present volume has been written.

The actual plants that lived and died long ago have left either no trace of their form and character, or but imperfect fragments of some of their parts embedded in hard rock and often hidden deep in the earth. That such difficulties lie in our way should not discourage us from attempting to learn all the fossils can teach. Many an old manuscript which is torn and partly destroyed bears a record, the fragments of which are more interesting and important than a tale told by a complete new book. The very difficulty of the subject of fossil botany is in itself an incentive to study, and the obstacles to be surmounted before a view of the ancient plants can be seen increase the fascination of the journey.

The world of to-day has been nearly explored; but the world, or rather the innumerable world-phases of the past, lie before us practically unknown, bewilderingly enticing in their mystery. These untrodden regions are revealed to us only by the fossils lying scattered through the rocks at our feet, which give us the clues to guide us along an adventurous path.

Fables of flying dragons and wondrous sea monsters have been shown by the students of animal fossils to be no more marvellous than were the actual creatures which once inhabited the globe; and among the plants such wonderful monsters have their parallels in the floras of the past. The trees which are living to-day are very recent in comparison with the ancestors of the families of lowlier plants, and most of the modern forest trees have usurped a position which once belonged to the monster members of such families as the Lycopods and Equisetums, which are now humble and dwindling. An ancient giant of the past is seen in the frontispiece, and the great girth of its stem offers a striking contrast to the feeble trailing branches of its living relatives, the Club-mosses

As we follow their histories we shall see how family after family has risen to dominate the forest, and has in its turn given place to a succeeding group. Some of the families that flourished long since have living descendants of dwarfed and puny growth, others have died out completely, so that their very existence would have been unsuspected had it not been revealed by their broken fragments entombed in the rocks.

From the study of the fossils, also, we can discover something of the course of the evolution of the different parts of the plant body, from the changes it has passed through in the countless ages of its existence. Just as the dominant animals of the past had bodies lacking in many of the characters which are most important to the living animals, so did the early plants differ from those around us to-day. It is the comparative study of living and fossil structures which throws the strongest light on the facts and factors of evolution.

When the study of fossil organisms goes into minute detail and embraces the fine subtleties of their internal structure, then the student of fossil plants has the advantage of the zoological observer, for in many of the fossil plants the cells themselves are petrified with a perfection that no fossil animal tissues have yet been found to approach. Under the microscope the most delicate of plant cells, the patterns on their walls, and sometimes even their nuclei can be recognized as clearly as if they were living tissues. The value of this is immense, because the external appearance of leaves and stems is often very deceptive, and only when both external appearance and internal structure are known can a real estimate of the character of the plant be made. In the following chapters a number of photographs taken through the microscope will show some of the cell structure from fossil plants. Such figures as fig. 11 and fig. 96, for example, illustrate the excellence of preservation which is often found in petrified plant tissues. Indeed, the microscope becomes an essential part of the equipment of a fossil botanist; as it is to a student of living plants. But for those who are not intending to specialize on the subject micro-photographs will illustrate sufficient detail, while in most modern museums some excellently preserved specimens are exhibited which show their structure if examined with a magnifying glass.

We recognize to-day the effect the vegetation of a district has on its scenery, even on its more fundamental nature; and we see how the plants keep in close harmony with the lands and waters, the climates and soils of the places they inhabit. So was it in the past. Hence the fossil plants of a district will throw much light on its physical characters during the epoch when they were living, and from their evidence it is possible to build up a picture of the conditions of a region during the epochs of its unwritten history.

From every point of view a student of living plants will find his knowledge and understanding of them greatly increased by a study of the fossils. Not only to the botanist is the subject of value, the geologist is equally concerned with it, though from a slightly different viewpoint, and all students of the past history of the earth will gain from it a wider knowledge of their specialty.

To all observers of life, to all philosophers, the whole history of plants, which only approaches completion when the fossils are studied, and compared or contrasted with living forms, affords a wonderful illustration of the laws of evolution on which are based most of the modern conceptions of life. Even to those whose profession necessitates purely practical lines of thought, fossil botany has something to teach; the study of coal, for instance, comes within its boundaries. While to all who think on the world at all, the story told by the fossil plants is a chapter in the Book of Life which is as well worth reading as any in that mystical volume.

CHAPTER II

VARIOUS KINDS OF FOSSIL PLANTS

Of the rocks which form the solid earth of to-day, a very large proportion have been built up from the deposits at the bottom of ancient oceans and lakes. The earth is very old, and in the course of its history dry land and sea, mountains and valleys have been formed and again destroyed on the same spot, and it is from the silt at the bottom of an ocean that the hills of the future are built.

The chief key we have to the processes that were in operation in the past is the course of events passing under our eyes to-day. Hence, if we would understand the formation of the rocks in the ancient seas, we must go to the shores of the modern ones and see what is taking place there. One of the most noticeable characters of a shore is the line of flotsam that is left by the edge of the waves; here you may find all kinds of land plants mixed with the sea shells and general rubbish, plants that may have drifted far. Much of the débris (outside towns) is brought down by the rivers, and may be carried some distance out to sea; then part becomes waterlogged and sinks, and part floats in to shore, perhaps to be carried out again, or to be buried under the coarse sand of the beach. When we examine sandstone rock, or the finer grained stones which are hardened mud, we find in them the remains of shells, sometimes of bones, and also of plant leaves and stems, which in their time had formed the flotsam of a shore. Indeed, one may say that nearly every rock which has not been formed in ancient volcanoes, or been altered by their heat, carries in it *some* trace of plant or animal. These remains are often very fragmentary and difficult to recognize, but sometimes they are wellnigh as perfect as dried specimens of living things. When they are recog-

VARIOUS KINDS OF FOSSIL PLANTS

nizable as plant or animal remains they are commonly called "fossils", and it is from their testimony that we must learn all we can know about the life of the past.

If we would find such stones for ourselves, the quarries offer the best hunting ground, for there several layers of rock are exposed, and we can reach fresh surfaces which have not been decayed by rain and storm.

Fig. 1.—The Face of a Quarry, showing layers or "beds" of different rock, *a, b,* and *c.* The top gravel and soil *s* has been disintegrated by the growing plants and atmosphere.

Fig. 1 shows a diagram of a quarry, and illustrates the almost universal fact that the beds of rock when undisturbed lie parallel to each other. Rock *a* in the figure is fine-grained limestone, *b* black friable shale mixed with sand, and *c* purer shale. In such a series of rocks the best fossils will be found in the limestone; its harder and finer structure acting as a better preservative of organisms than the others. In limestone one finds both plant and animal fossils, very often mixed together as the flotsam on the shore is mixed. Many limestones split along parallel planes, and may break into quite thin sheets on whose surfaces the flattened fossils show particularly well.

It is, however, with the plant fossils that we must concern ourselves, and among them we find great variety of form. Some are more or less complete, and give an immediate idea of the size and appearance of the plant to which they had belonged; but such are rare. One of the best-known examples of this type is the base of a great tree trunk illustrated in the frontispiece. With such a fossil there is no shadow of doubt that it is part of a giant tree, and its spreading roots running so far horizontally along the ground suggest the picture of a large crown of branches. Most fossils, however, are much less illuminating, and it is usually only by the careful piecing together of fragments that we can obtain a mental picture of a fossil plant.

A fossil such as that illustrated in the frontispiece—and on a smaller scale this type of preservation is one of the commonest—does not actually consist of the plant body itself. Although from the outside it looks as though it were a stem base covered with bark, the whole of the inner portion is composed of fine hard rock with no trace of woody tissue. In such specimens we have the shape, size, and form of the plant preserved, but none of its actual structure or cells. It is, in fact, a CAST. Fossil casts appear to have been formed by fine sand or mud silting round a submerged stump and enclosing it as completely as if it had been set in plaster of Paris; then the wood and soft tissue decayed and the hollow was filled up with more fine silt; gradually all the bark also decayed and the mud hardened into stone. Thus the stone mould round the outside of the plant enclosed a stone casting. When, after lying for ages undisturbed, these fossils are unearthed, they are so hard and "set" that the surrounding stone peels away from the inner part, just as a plaster cast comes away from an object and retains its shape. There are many varieties of casts among fossil plants. Sometimes on breaking

VARIOUS KINDS OF FOSSIL PLANTS

a rock it will split so as to show the perfect form of the surface of a stem, while its reverse is left on the stone as is shown in fig. 2. Had we only the reverse we should still have been able to see the form of the leaf bases by taking a wax impression from it; although there is nothing of the actual tissue of the plant in such a fossil. Sometimes casts of leaf bases

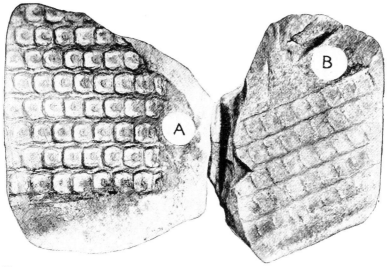

Fig. 2.—A, Cast of the Surface showing the Shape of Leaf Bases of *Sigillaria*; B, the reverse of the impression left on the adjacent layer of rock. (Photo.)

show the detail preserved with wonderful sharpness, as in fig. 3. This is an illustration of the leaf scars of *Lepidodendron*, which often form particularly good casts.

In other instances the cast may simply represent the internal hollows of the plant. This happens most commonly in the case of stems which contained soft pith cells which quickly decayed, or with naturally hollow stems like the Horse-tails (*Equisetum*) of to-day. Fine mud or sand silted into such hollows completely filling them up, and then, whether the rest of the plant were preserved or not, the shape of the inside of the

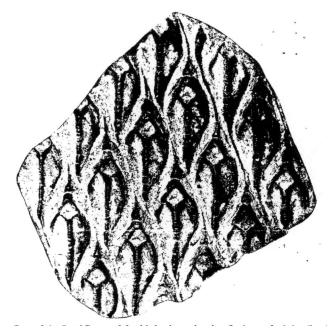

Fig. 3.—Cast of the Leaf Bases of *Lepidodendron*, showing finely marked detail. (Photo.)

stem remains as a solid stone. Where this has happened, and the outer part of the plant has decayed so as to leave no trace, the solid plug of stone from the centre may look very much like an actual stem itself, as it is cylindrical and may have surface markings like those on the outsides of stems. Some of the casts of this type were for long a puzzle to the older fossil botanists, particularly that illustrated in fig. 4, where the whole looks like a pile of discs.

The true nature of this fossil was recognized when casts of the plant were found with some of the wood preserved outside the castings; and

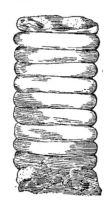

Fig. 4.—"*Sternbergia*." Internal cast of the stem of *Cordaites*.

VARIOUS KINDS OF FOSSIL PLANTS

it was then known that the plant had a hollow pith, with transverse bands of tissue across it at intervals which caused the curious constrictions in the cast.

Another form of cast which is common in some rocks is that of seeds. As a rule these casts are not connected with any actually preserved tissue, but they

Fig. 5.—Leaf Impressions of "Fern" *Sphenopteris* on Shale. (Photo.)

show the external form, or the form of the stony part of the seed. Well-known seeds of this type are those of *Trigonocarpon*, which has three characteristic ridges down the stone. Sometimes in the fine sandstone in which they occur embedded, the *internal* cast lies embedded *in the external* cast, and between them there is a slight space, now empty, but which once contained the actual shell of the seed, now decayed.

Thus we may rattle the "stone" of a fossil fruit as we do the dried nuts of to-day—the external resemblance between the living and the fossil is very striking, but of the actual tissues of the fossil seed nothing is left.

Casts have been of great service to the fossil botanists, for they often give clear indications of the external appearance of the parts they represent; particularly of stems, leaf scars, and large seeds. But all such fossils are very imperfect records of the past plants, for none of the actual plant tissues, no minute anatomy or cell structure, is preserved in that way.

A type of fossil which often shows more detail, and which usually retains something of the actual tissues of the plant, is that known technically as the IMPRESSION. These fossils are the most attractive of all the many kinds we have scattered through the rocks, for they often show with marvellous perfection the most delicate and beautiful fern leaves, such as in fig. 5. Here the plant shows up as a black silhouette against the grey stone, and the very veins of the midrib and leaves are quite visible.

Fig. 6 shows another fernlike leaf in an impression, not quite flat like that shown in fig. 5, but with a slight natural curvature of the leaves similar to what would have been their form in life. Though an impression, this specimen is not of the "pressed plant" type, it almost might be described as a *bas-relief*.

Sometimes impressions of fern foliage are very large, and show highly branched and complex leaves like those of tree ferns, and they may cover large sheets of stone. They are particularly common in the fine shales above coal seams, and are best seen in the mines, for they are often too big to bring to the surface complete.

In most impressions the black colour is due to a film of carbon which represents the partly decomposed tissues of the plant. Sometimes this film is cohesive enough to be detached from the stone without damage.

Fig. 6.—Impression of *Neuropteris* Leaf, showing details of veins, the leaves in partial relief. (Photo.)

Beautiful specimens of this kind are to be seen in the Royal Scottish Museum, Edinburgh where the coiled bud of a young fern leaf has been separated from the rock on which it was pressed, and mounted on glass. Such specimens might be called mummy plants, for they are the actual plant material, but so decayed and withered that the internal cells are no longer intact. In really well preserved ones it is sometimes possible to peel off the plant film, and then treat it with strong chemical agents to clear the black carbon atoms away, and mount it for microscopic examination, when the actual outline of the epidermis cells can be seen.

In fig. 7, the impression is that of a *Ginkgo* leaf, and after treatment the cells of the epidermis were perfectly recognizable under the microscope, with the stomates (breathing pores) also well preserved. This is shown in fig. 8, where the outline of the cells was drawn from the microscope. In such specimens, however, it is only the outer skin which is preserved, the inner soft tissue, the vital anatomy of the plant, is crushed and carbonized.

Fig. 7.—Leaf Impression of *Ginkgo*, of which the film was strong enough to peel off complete

Leaves, stems, roots, even flowers (in the more recent rocks) and seeds may all be preserved as impressions; and very often those from the more recently formed rocks are so sharply defined and perfect that they seem to be actual dried leaves laid on the stone.

Much evidence has been accumulated that goes to show that the rocks which contain the best impressions were originally deposited under tranquil conditions in water. It might have been in a pool or quiet lake with overshadowing trees, or a landlocked inlet of the sea where silt quietly accumulated, and as the plant fragments fell or drifted into the spot they were covered by fine-grained mud without disturbance. In

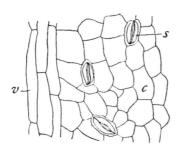

Fig. 8.—Outline of the Cells from Specimen of Leaf shown in fig. 7
c, Ordinary cells; *s*, stomates; *v*, elongated cells above the vein.

the case of those which are very well preserved this must have taken place with considerable rapidity, so that they were shut away from contact with the air and from the decay which it induces.

Impressions in the thin sheets of fine rock may be compared to dried specimens pressed between sheets of blotting paper; they are flattened, preserved from decay, and their detailed outline is retained. Fossils of this kind are most valuable, for they give a clear picture of the form of the foliage, and when, as sometimes happens, large masses of leaves, or branches with several leaves attached to them, are preserved together, it is possible to reconstruct the plant from them. It is chiefly from such impressions that the inspiration is drawn for those semi-imaginary pictures of the forests of long ago. From them also are drawn many facts of prime importance to scientists about the nature and appearance of plants, of which the internal anatomy is known from other specimens, and also about the connection of various parts with each other.

Sometimes isolated impressions are found in clay balls or nodules. When the latter are split open they may show as a centre or nucleus a leaf or cone, round which the nodule has collected. In such cases the plant is often preserved without compression, and may show something of the minute details of organization. The preservation, however, is generally far from perfect when viewed from a microscopical standpoint. Fig. 9 shows one of these smooth, clayey nodules split open, and within it the cone which formed its centre, also split into two, and standing in high relief, with its scales showing clearly. Similar nodules or balls of clay are found to-day, forming in slowly running water, and it may be generally observed that they collect round some rubbish, shell, or plant fragment. These nodules are particularly well seen nowadays in the mouth of the Clyde, where they are formed with great rapidity.

16 ANCIENT PLANTS

Another kind of preservation is that which coats over the whole plant surface with mineral matter, which hardens, and thus preserves the *form* of the plant. This process can be observed going on to-day in the neighbourhood of hot volcanic streams where the water is heavily charged with minerals. In most cases such

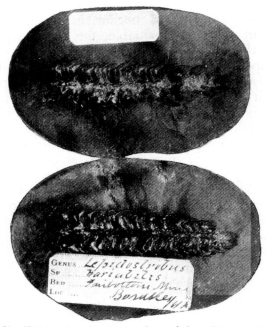

Fig. 9.—Clay Nodule split open, showing the two halves of the cone which was its centre. (Photo.)

fossils have proved of little importance to science, though there are some interesting specimens in the French museums which have not yet been fully examined. A noteworthy fossil of this type is the *Chara*, which, growing in masses together, has sometimes been preserved in this way in large quantities, indicating the existence of an ancient pond in the locality.

There is quite a variety of other types of preser-

VARIOUS KINDS OF FOSSIL PLANTS

vation among fossil plants, but they are of minor interest and importance, and hardly justify detailed consideration. One example that should be mentioned is Amber. This is the gum of old resinous trees, and is a well-known substance which may rank as a "fossil". Jet, too, is formed from plants, while coal is so important that the whole of the next chapter will be devoted to its consideration. Even the black lead of pencils possibly represents plants that were once alive on this globe.

Though such remains tell us of the existence of plants at the place they were found at a known period in the past, yet they tell very little about the actual structure of the plants themselves, and therefore very little that is of real use to the botanist. Fortunately, however, there are fossils which preserve every cell of the plant tissues, each one perfect, distended as in life, and yet replaced by stone so as to be hard and to allow of the preparation of thin sections which can be studied with the microscope. These are the vegetable fossils which are of prime importance to the botanist and the scientific enquirer into the evolution of plants. Such specimens are commonly known as PETRIFACTIONS.

Sometimes small isolated stumps of wood or branches have been completely permeated by silica, which replaces the cell walls and completely preserves and hardens the tissues. This silicified wood is found in a number of different beds of rock, and may be seen washed out on the shore in Yorkshire, Sutherland, and other places where such rocks occur. When such a block is cut and polished the annual rings and all the fine structure or "grain" of the wood become as apparent as in recent wood. From these fossils, too, microscopic sections can be cut, and then the individual wood cells can be studied almost as well as those of living trees. A particularly notable example of fossil tree trunks is the Tertiary forest of the Yellowstone Park. Here the petrified trunks are weathered out and stand together much as they must have stood when alive; they are of course bereft of their foliage branches.

Such specimens, however, are usually only isolated blocks of wood, often fragments from large stumps which show nothing but the rings of late-formed wood. It is impossible to connect them with the impressions of leaves or fruits in most cases, so that of the plants they represent we know only the anatomical structure of the secondary wood and nothing of the foliage or general appearance of the plant as a whole. Hence these specimens also give a very partial representation of the plants to which they belonged.

Fortunately, however, there is still another type of preservation of fossils, a type more perfect than any of the others and sometimes combining the advantages of all of them. This is the special type of petrifaction which includes, not a single piece of wood, but a whole mass of vegetation consisting of fragments of stems, roots, leaves, and even seeds, sometimes all together. These petrifactions are those of masses of forest débris which were lying as they dropped from the trees, or had drifted together as such fragments do. The plant tissues in such masses are preserved so that the most delicate soft tissue cells are perfect, and in many cases the sections are so distinct that one might well be deluded into the belief that it is a living plant at which one looks.

Very important and well-known specimens have been found in France and described by the French palæobotanists. As a rule these specimens are preserved in silica, and are found now in irregular masses of the nature of chert. Of still greater importance, however, owing partly to their greater abundance and partly to the quantity of scientific work that has been done on them, are the masses of stone found in the English coal seams and commonly called "coal balls".

The "coal balls" are best known from Lancashire

VARIOUS KINDS OF FOSSIL PLANTS 19

and Yorkshire, where they are extremely common in some of the mines, but they also occur in Westphalia and other places on the Continent.

In external appearance the "coal balls" are slightly irregular roundish masses, most generally about the size of potatoes, and black on the outside from films of adhering coal. Their size varies greatly, and they have

Fig. 10.—Mass of Coal with many "coal balls" embedded in it

a a, In surface view; *b b*, cut across. All washed with acid to make the coal balls show up against the black coal. (Photo by Lomax.)

been found from that of peas up to masses with a diameter of a foot and a half. They lie embedded in the coal and are not very easily recognizable in it at first, because they are black also, but when washed with acid they turn greyish-white and then can be recognized clearly. Fig. 10 shows a block of coal with an exceptionally large number of the "coal balls" embedded in it. This figure illustrates their slightly irregular rounded form in a typical manner. By chemical analysis they are found to consist of a nearly pure mixture of the carbonates of lime and magnesia;

though in some specimens there is a considerable quantity of iron sulphide, and in all there is at least 5 per cent of various impurities and some quantity of carbon.

The important mineral compounds, $CaCO_3$ and $MgCO_3$, are mixed in very different quantities, and even

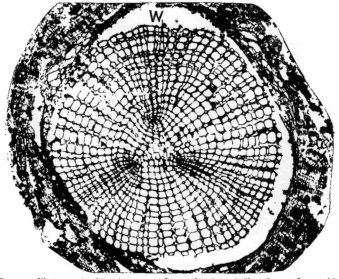

Fig. 11.—Photograph of Section across Stem of *Sphenophyllum* from a Lancashire "coal ball", showing perfect preservation of woody tissue

w, wood; c, cortex.

in coal balls lying quite close to each other there is often much dissimilarity in this respect. In whatever proportion these minerals are combined, it seems to make but little difference to their preservative power, and in good "coal balls" they may completely replace and petrify each individual cell of the plants in them.

Fig. 11 shows a section across the wood of a stem preserved in a "coal ball", and illustrates a degree of perfection which is not uncommon. In the course of the succeeding chapters constant reference will be made to tissues preserved in "coal balls", and it may be noticed that not only the relatively hard woody cells are preserved but the very softest and youngest tissues also appear equally unharmed by their long sojourn in the rocks.

Fig. 12.—Photograph of Section through a Bud of *Lepidodendron*, showing many small leaves tightly packed round the axis. From a "coal ball"

The particular value of the coal balls as records of past vegetation lies in the fact that they are petrifactions, not of individual plants alone, but of masses of plant débris. Hence in one of these stony concretions may lie twigs with leaves attached, bits of stems with their fruits, and fine rootlets growing through the mass. A careful study and comparison of these fragments has led to the connection, piece by piece, of the various parts of many plants. Such a specimen as that figured in fig. 12 shows how the soft tissues of young leaves are preserved, and how their relation to each other and to the axis is indicated.

Hitherto the only concretions of the nature of "coal balls" containing well preserved plant débris, have been found in the coal or immediately above it, and are of Palæozoic age (see p. 34). Recent exploration, however, has resulted in the discovery of similar concretions of Mesozoic age, from which much may be hoped in the future. Still, at present, it is to the palæozoic specimens we must turn for nearly all valuable knowledge about ancient plants, and primarily to that form of preservation of the specimens known as structural petrifactions, of which the "coal balls" are both the commonest and the most perfect examples.

CHAPTER III

COAL, THE MOST IMPORTANT OF PLANT REMAINS

Some of the many forms which are taken by fossil plants were shortly described in the last chapter, but the most important of all, namely coal, must now be considered. Of the fossils hitherto mentioned many are difficult to recognize without examining them very closely, and one might say that all have but little influence on human life, for they are of little practical or commercial use, and their scientific value is not yet very widely known. Of all fossil plants, the great exception is coal. Its commercial importance all over the world needs no illustration, and its appearance needs no description for it is in use in nearly every household. Quite apart from its economic importance, coal has a unique place among fossils in the eyes of the scientist, and is of special interest to the palæontologist.

In England nearly all the coal lies in rocks of a

great age, belonging to a period very remote in the world's history. The rocks bearing the coal contain other fossils, principally those of marine animals, which are characteristic of them and of the period during which they were formed, which is generally known as the "Coal Measure period". There is geological proof that at one time the coal seams were much more widely spread over England than they are at present; they have been broken up and destroyed in the course of ages, by the natural movements among the rocks and by the many changes and processes of disintegration and decay which have gone on ever since they were deposited. To-day there are but relatively small coal-bearing areas, which have been preserved in the hollows of the synclines.[1]

The seams of coal are extremely numerous, and even the same seam may vary greatly in thickness. From a quarter of an inch to five or six feet is the commonest thickness for coal in this country, but there are many beds abroad of very much greater size. Thin seams often lie irregularly in coarse sandstone; for example, they may be commonly seen in the Millstone Grit; but typical coal seams are found embedded between rocks of a more or less definite character known as the "roof" and "floor".

Basalts, granites, and such rocks do not contain coal; the coal measures in which the seams of coal occur are, generally speaking, limestones, fine sandstones, and shales, that is to say, rocks which in their origin were deposited under water. In detail almost every seam has some individual peculiarity, but the following represents two types of typical seams. In many cases, below the coal, the limestone or sandstone rocks give place to fine, yellow-coloured layers of clay, which varies from a

[1] The student would do well to read up the general geology of this very interesting subject. Such books as Lyell's *Principles of Geology*, Geikie's textbooks, and many others, provide information about the process of "mountain building" on which the form of our coalfields depends. A good elementary account is to be found in Watt's *Geology for Beginners*, p. 96 *et seq*.

few inches to many feet in thickness and is called the "underclay". This fine clay is generally free from pebbles and coarse débris of all kinds, and is often supposed to be the soil in which the plants forming the coal had been growing. The line of demarcation between the coal and the clay is usually very sharp, and the compact black layers of hard coal stop almost as abruptly on the upper side and give place to a shale or

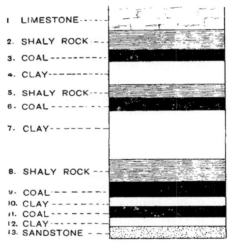

Fig. 13.—Diagram of a Series of Parallel Coal Seams with Underclays and Shale Roofs of varying thicknesses

limestone "roof"; see fig. 13, layers 5, 6, and 7. Very frequently a number of small seams come together, lying parallel, and sometimes succeeding each other so rapidly that the "roof" is eliminated, and a clay floor followed by a coal seam, is succeeded immediately by another clay floor and another coal seam, as in fig. 13, layers 10, 11, and 12. The relative thickness of these beds also varies very greatly, and over an underclay of seven or eight feet the coal seam may only reach a couple of inches, while a thick seam may have a floor of very slight dimensions. These relations depend on

COAL

such a variety of local circumstances from the day they were forming, that it is only possible to unravel the causes when an individual case is closely studied. The main sequence, however, is constant and is that illustrated in fig. 13.

The second type of seam is that in which the underclay floor is not present, and is replaced either by shales or by a special very hard rock of a finely granular nature called "gannister". In the gannister floor it is usual to find traces of rootlets and basal stumps of plants, which seem to indicate that the gannister was the

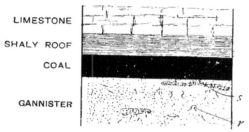

Fig. 14.—Diagram of Coal Seam with Gannister Floor, in which are traces of rootlets r, and of stumps of root-like organs s

ground in which the plants forming the coal were rooted. The coal itself is generally very pure plant remains, though between its layers are often found bands of shaly stone which are called "dirt bands". These are particularly noticeable in thick seams, and they may be looked on as corresponding to the roof shales; as though, in fact, the roof had started to form but had only reached a slight development when the coal formation began again.

That the coal is strikingly different from the rocks in which it lies is very obvious, but that alone is no indication of its origin. It is now so universally known and accepted that coal is the remains of vegetables that no proofs are usually offered for the statement. It is, however, of both interest and importance to marshal the evidence for this belief. The grounds for recognizing coal as consisting of practically pure plant remains are many and various, so that only the more important of them will be considered now. The most direct suggestion lies in the impressions of leaves and stems which are found between its layers; this, however, is confronted by the parallel case of plant impressions found in shales and limestones which are not of vegetable origin, so that it might be argued that those plants in the coal drifted in as did those in the limestone. But when we examine the black impressions on limestone or sandstone, an item of value is noticeable; it is often possible to peel off a film, lying between the upper and lower impression, of black coaly substance, sometimes an eighth of an inch thick, and hard and shining like coal. This follows the outline of the plant form of the impression, and it is certain that this minute "coal seam" was formed from the plant tissues. It is, in fact, a coal seam bearing the clearest possible evidence of its plant nature. We have only to imagine this multiplied by many plants lying tightly packed together, with no mineral impurities between, to see that it would yield a coal seam like those we find actually existing.

In some cases in the coal itself a certain amount of the structure of the plants which formed it remains, though usually, in the process of their decay the tissues have entirely decomposed, and left only their carbonized elements. Chemical analysis reveals that, beyond the percentage of mineral ash which is found in living plants, there is little in a pure sample of coal that is not carbonaceous. All the deposits of carbon found in any form in nature can be traced to some animal or vegetable remains, so that it is logical to assume that coal also arose from either animal or plant débris. But were coal of an animal origin, the amount of mineral matter in it would be much larger as well as being of a different nature; for almost all animals have skeletons, even the simplest single-celled protozoa often own calcareous shells, sponges have siliceous spicules, molluscs hard shells, and the higher animals bones and teeth. These things are of a very permanent nature, and would certainly be found in quantities in the coal had animals formed it. Further, the peat of to-day, which collects in thick compact masses of vegetable, shows how plants may form a material consisting of carbonized remains.

Fig. 15.—Part of a Coal Ball, showing the concentric bandings in it which are characteristic of concretions

By certain experiments in which peat was subjected to pressure and heat, practically normal coal was made from it.

Still a further witness may be found in the structure of the "coal balls" described in the last chapter. These stony masses, lying in the pure coal, might well be considered as apart from it and bearing no relation to its structure; but recent work has shown that they were actually formed at the same time as the coal, developing in its mass as mineral concretions round some of the plants in the soft, saturated, peaty mass which was to be hardened into coal later on.[1] All "coal balls" do not

[1] See note on p. 28.

show their concretionary structure so clearly, but sometimes it can be seen that they are made with concentric bands or markings like those characteristic of ordinary mineral concretions (see fig. 15). Concretions are formed by the crystallization of minerals round some centre, and it must have happened that in the coal seams in which the coal-ball concretions are found that this process took place in the soft plant mass before it hardened. Recent research has found that there is good evidence that those

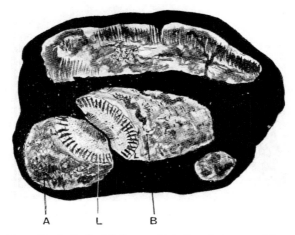

Fig. 16.—Mass of Coal with Coal Balls, A and B both enclosing part of the same stem L

seams[1] resulted from the slow accumulation of plant débris under the salt or brackish water in whose swamps the plants were growing, and that as they were collecting the ground slowly sank till they were quite below the level of the sea and were covered by marine silt. At the same time some of the minerals present in the sea water, which must have saturated the mass, crystallized partly and deposited themselves round centres in the plant tissues, and by enclosing them and penetrating

[1] This refers only to the "coal-ball"-bearing seams; there are many other coals which have certainly collected in other ways. See Stopes & Watson, Appendix, p. 187.

them preserved them from decay till the mineral structure entirely replaced the cells, molecule by molecule. Evidence is not wanting that this process went on without disturbance, for in fig. 16 is shown a mass of coal in which lie several coal balls, two of which enclose parts of the same plant. This means that round different centres in the same stem two of the concretions were forming and preserving the tissues; the two stone masses, however, did not enlarge enough to unite, but left a part of the tissue unmineralized, which is now seen as a streak of coal. We have here the most important proof that the coal balls are actually formed in the coal and of the plants making the coal, for had those coal balls come in as pebbles, or in any way from the outside into the coal, they could not have remained in such a position as to lie side by side enclosing part of the same stem. There are many other details which may be used in this proof, but this one illustration serves to show the importance of coal balls when dealing with the theories of the origin of coal, for they are perfectly preserved samples of what the whole coal mass was at one time.

There are but few seams, however, which contain coal balls, and about those in which they do not occur our knowledge is very scanty. It is often assumed that the plant impressions in the shales above the coal seams can be taken as fair samples of those which formed the coal itself; but this has been recently shown to be a fallacious argument in some cases, so that it is impossible to rely on it in general. The truth is, that though coal is one of the most studied of all the geological deposits, we are still profoundly ignorant of the details of its formation except in a few cases.

The way in which coal seams were formed has been described often and variously, and for many years there were heated discussions between the upholders of the different views as to the merits of their various theories. It is now certain that there must have been at least four principal ways in which coal was formed, and the different seams are illustrations of the products of different methods. In all cases more or less water is required, for coal is what is known as a sedimentary deposit, that is, one which collects under water, like the fine mud and silt and débris in a lake. It will be understood, however, that if the plant remains were collecting at any spot, and the water brought in sand and mud as well, then the deposit could not have resulted in pure coal, but would have been a sandy mixture with many plant remains, and would have resulted in the formation of a rock, such as parts of the millstone grit, where there are many streaks of coal through the stone.

Among various coal seams, evidence for the following modes of coal formation can be found:—

(*a*) *In fresh water.*—In still freshwater lakes or pools, with overhanging plants growing on the banks, twigs and leaves which fell or were blown into the water became waterlogged and sank to the bottom. With a luxuriant growth of plants rapidly collecting under water, and there preserved from contact with the air and its decaying influence, enough plant remains would collect to form a seam. After that some change in the local conditions took place, and other deposits covered the plants and began the accumulations which finally pressed the vegetable mass into coal.

To freshwater lakes of large size plants might also have been brought by rivers and streams; they would have become waterlogged in time, after floating farther than the sand and stones with which they came, and would thus settle and form a deposit practically free from anything but plant remains.

(*b*) *As peat.*—Peat commonly forms on our heather moors and bogs to-day to a considerable thickness. This also took place long ago in all probability, and when the level of the land altered it would have been covered by other deposits, pressed, and finally changed into coal.

(*c*) *In salt or brackish water, growing in situ.*—Trees

and undergrowth growing thickly together in a salt or brackish marsh supplied a large quantity of débris which fell into the mud or water below them, and were thus shut off from the air and partly preserved. When conditions favoured the formation of a coal seam the land level was slowly sinking, and so, though the débris collected in large quantities, it was always kept just beneath the water level. Finally the land sank more rapidly, till the vegetable mass was quite under sea water, then mud was deposited over it, and the materials which were afterwards hardened to form the roof rocks were deposited. This was the case in those seams in which "coal balls" occur, and the evidence of the sea water covering the coal soon after it was deposited lies in the numerous sea shells found in the roof immediately above it.

(*d*) *In salt water, drifted material.*—Tree trunks and large tangled masses of vegetation drifted out to sea by the rivers just as they do to-day. These became waterlogged, and finally sank some distance from the shore. (Those sinking near the shore would not form pure coal, for sand and mud would be mixed with them, also brought down by rivers and stirred up from the bottom by waves.) The currents would bring numbers of such plants to the same area until a large mass was deposited on the sea floor. Finally the local conditions would have changed, the currents then bringing mud or sand, which covered the vegetable mass and formed the mineral roof of the resulting coal seam. There is a variety of what might be called the "drifted coals", which appears to have been formed of nothing but the *spores* of plants of a resinous nature. These structures must have been very light, and possibly floated a long distance before sinking.

If we could but obtain enough evidence to understand each case fully we should probably find that every coal seam represents some slightly different mode of formation, that in each case there was some local peculiarity in the plants themselves and the way they accumu lated in coal-forming masses, but the above four methods will be found to cover the principal ways in which coal has arisen.

Coal, as we now know it, has a great variety of qualities. The differences probably depend only to a small extent on the varieties among the plants forming it, and are almost entirely due to the many later conditions which have affected the coal after its original formation. Some such conditions are the various upheavals and depressions to which the rocks containing the coal have been subjected, the weight of the beds lying over the coal seams, and the high temperatures to which they may have been subjected when lying under a considerable depth of later-deposited rocks. The influence on the coal of these and many other physical factors has been enormous, but they are purely cosmical and belong to the special realm of geological study, and so cannot be considered in detail now.

To return to our special subject, namely, the plants themselves which are now preserved in the coal. Their nature and appearance, their affinities and minute structure, can only be ascertained by a detailed study, to which the following chapters will be devoted, though in their limited space but an outline sketch of the subject can be drawn.

It has been stated by some writers that in the Coal Measure period plants were more numerous and luxuriant than they ever were before or ever have been since. This view could only have been brought forward by one who was considering the geology of England alone, and in any case there appears to be very little real evidence for such a view. Certainly in Europe a large proportion of the coal is of this age, and to supply the enormous masses of vegetation it represents a great growth of plants must have existed. But it is evident that just at the Carboniferous period in what is now called Europe the physical conditions of the land which roughly corresponded to the present Continent were such as favoured the accumulation of plants, and the gradual sinking of the land level also favoured their preservation under rapidly succeeding deposits. Of the countless plants growing in Europe to-day very few stand any chance of being preserved as coal for the future; so that, unless the physical conditions were suitable, plants might have been growing in great quantity at any given period without ever forming coal. But now that the geology of the whole world is becoming better known, it is found that coal is by no means specially confined to the Coal Measure age. Even in Europe coals of a much later date are worked, while abroad, especially in Asia and Australia, the later coals are very important. For example, in Japan, seams of coal 14, 20, and even more feet in thickness are worked which belong to the Tertiary period (see p. 34), while in Manchuria coal 100 feet thick is reported of the same age. When these facts are considered it is soon found that all the statements made about the unique vegetative luxuriance of the Coal Measure period are founded either on insufficient evidence or on no evidence at all.

The plants forming the later coals must have had in their own structure much that differed from those forming the old coals of Britain, and the gradual change in the character of the vegetation in the course of the succeeding ages is a point of first-rate importance and interest which will be considered shortly in the next chapter.

CHAPTER IV

THE SEVEN AGES OF PLANT LIFE

Life has played its important part on the earth for countless series of years, of the length of whose periods no one has any exact knowledge. Many guesses have been made, and many scientific theories have been used to estimate their duration, but they remain inscrutable. When numbers are immense they cease to hold any meaning for us, for the human mind cannot comprehend the significance of vast numbers, of immense space, or of æons of time. Hence when we look back on the history of the world we cannot attempt to give even approximate dates for its events, and the best we can do is to speak only of great periods as units whose relative position and whose relative duration we can estimate to some extent.

Those who have studied geology, which is the science of the world's history since its beginning, have given names to the great epochs and to their chief subdivisions. With the smaller periods and the subdivisions of the greater ones we will not concern ourselves, for our study of the plants it will suffice if we recognize the main sequence of past time.

The main divisions are practically universal, and evidence of their existence and of the character of the creatures living in them can be found all over the world; the smaller divisions, however, may often be local, or only of value in one continent. To the specialist even the smallest of them is of importance, and is a link in the chain of evidence with which he cannot dispense; but we are at present concerned only with the broad outlines of the history of the plants of these periods, so will not trouble ourselves with unnecessary details.[1] Corresponding to certain marked changes in the character of the vegetation, we find seven important divisions of geological time which we will take as our unit periods, and which are tabulated as follows:—

Cainozoic { I. Present Day.
{ II. Tertiary.
Mesozoic { III. Upper Cretaceous (or Chalk).
{ IV. The rest of the Mesozoic.
Palæozoic { V. Newer Palæozoic, including { Permian. Carboniferous. Devonian.
{ VI. Older Palæozoic.
Eozoic ... VII. Archæan.

[1] For a detailed list of the strata refer to Watts, p. 219 (see Appendix).

Now the actual length of these various periods was very different. The epoch of the Present Day is only in its commencement, and is like a thin line if compared with the broad bands of the past epochs. By far the greatest of the periods is the Archæan, and even the Older Palæozoic is probably longer than all the others taken together. It is, however, so remote, and the rocks which were formed in it retain so little plant structure that is decipherable, so few specimens which are more than mere fragments, that we know very little about it from the point of view of the plant life of the time. It includes the immense indefinite epochs when plants began to evolve, and the later ones when animals of many kinds flourished, and when plants, too, were of great size and importance, though we are ignorant of their structure. Of all the seven divisions of time, we can say least about the two earliest, simply for want of anything to say which is founded on fact rather than on theoretical conclusions.

Although these periods seem clearly marked off from one another when looked at from a great distance, they are, of course, but arbitrary divisions of one long, continuous series of slow changes. It is not in the way of nature to make an abrupt change and suddenly shut off one period—be it a day or an æon—from another, and just as the seasons glide almost imperceptibly into one another, so did the great periods of the past. Thus, though there is a strong and very evident contrast between the plants typical of the Carboniferous period and of the Mesozoic, those of the Permian are to some extent intermediate, and between the beginning of the Permian and the end of the Carboniferous—if judged by the flora—it is often hard to decide.

It must be realized that almost any given spot of land—the north of England, for example—has been beneath the sea, and again elevated into the air, at least more than once. That the hard rocks which make its present-day hills have been built up from the silt and débris under an ocean, and after being formed have seen daylight on a land surface long ago, and sunk again to be covered by newer deposits, perhaps even a second or a third time, before they rose for the time that is the present. Yet all these profound changes took place so slowly that had we been living then we could have felt no motion, just as we feel no motion to-day, though the land is continuing to change all around us. The great alternations between land and water over large areas mark out to some extent the main periods tabulated on p. 34, for after each great submersion the rising land seems to have harboured plants and animals with somewhat different characters from those which inhabited it before. Similarly, when the next submersion laid down more rocks of limestone and sandstone, they enclosed the shells of some creatures different from those which had inhabited the seas of the region previously.

Through all the periods the actual rocks formed are very similar—shales, limestones, sandstones, clays. When any rocks happen to have preserved neither plant nor animal remains it is almost impossible to tell to which epoch they belong, except from a comparative study of their position as regards other rocks which do retain fossils. This depends on the fact that the physical processes of rock building have gone on throughout the history of the globe on very much the same lines as they are following at present. By the sifting power of water, fine mud, sand, pebbles, and other débris are separated from each other and collected in masses like to like. The fine mud will harden into shales, sand-grains massed together harden into sandstones, and so on, and when, after being raised once more to form dry land, they are broken up by wind and rain and brought down again to the sea, they settle out once again in a similar way and form new shales and sandstones; and so on indefinitely. But meantime the living things, both plant and animal, have been changing, growing, evolving, and the leafy twig brought down with the sandgrains in the flooded river of one epoch differs from that brought down by the river of a succeeding epoch—though it might chance that the sandgrains were the same identical ones. And hence it is by the remains of the plants and animals in a rock that we can tell to which epoch it belonged. Unless, of course, ready-formed fossils from an earlier epoch get mixed with it, coming as pebbles in the river in flood—but that is a subtle point of geological importance which we cannot consider here. Such cases are almost always recognizable, and do not affect the main proposition.

From the various epochs, the plants which have been preserved as fossils are in nearly all cases those which had lived on the land, or at least on swamps and marshes by the land. Of water plants in the wide sense, including both those growing in fresh water and those in the sea, we have comparatively few. This lack is particularly remarkable in the case of the seaweeds, because they were actually growing in the very medium in which the bulk of the rocks were formed, and which we know from recent experiments acts as a preservative for the tissues of land plants submerged in it. It must be remembered, however, that almost all the plants growing in water have very soft tissues, and are usually of small size and delicate structure as compared with land plants, and thus would stand less chance of being preserved, and would also stand less chance of being recognized to-day were they preserved. The mark on a stone of the impression of a soft film of a waterweed would be very slight as compared with that left by a leathery leaf or the woody twig of a land plant.

There are, of course, exceptions, and, as will be noted later on (see Chapter XVII), there are fossil seaweeds and fossil freshwater plants, but we may take it on the whole that the fossils we shall have to deal with and that give important evidence, are those of the land which had drifted out to sea, in the many cases when they are found in rocks together with sea shells.

Let us now consider very shortly the salient features of the seven epochs we have named as the chief divisions of time. The vegetation of the CARBONIFEROUS PERIOD is better known to us than that of any other period except that of the present day, so that it will form the best starting-point for our consideration.

At this period there were, as there are to-day, oceans and continents, high lands, low lands, rivers and lakes, in fact, all the physical features of the present-day world, but they were all in different places from those of to-day. If we confine our attention to Britain, we find that at that period the far north, Scotland, Wales, and Charnwood were higher land, but the bulk of the southern area was covered by flat swamps or shallow inlets, where the land level gradually changed, slowly sinking in one place and slowly rising in others, which later began also to sink. Growing on this area wherever they could get a foothold were many plants, all different from any now living. Among them none bore flowers. A few families bore seeds in a peculiar way, differing widely from most seed-bearing plants of to-day. The most prevalent type of tree was that of which a stump is represented in the frontispiece, and of which there were many different species. These plants, though in size and some other ways similar to the great trees of to-day, were fundamentally different from them, and belonged to a very primitive family, of which but few and small representatives now exist, namely the Lycopods. Many other great trees were like hugely magnified "horsetails," or Equisetums; and there were also seed-bearing Gymnosperms of a type now extinct. There were ferns of many kinds, of which the principal ones belong to quite extinct families, as well as several other plants which have no parallel among living ones. Hence one may judge that the vegetation was rich and various, and that, as there were tall trees with seeds, the plants were already very highly evolved. Indeed, except for the highest group of all, the flowering plants, practically all

THE SEVEN AGES OF PLANT LIFE

the main groups now known were represented. The flora of the Devonian was very similar in essentials.

If that be so, it may seem unsatisfactory to place all the preceding æons under one heading, the OLDER PALÆOZOIC. And, indeed, it is very unsatisfactory to be forced to do so. We know from the study of animal fossils that this time was vast, and that there were several well-defined periods in it during which many groups of animals evolved, and became extinct after reaching their highest development; but of the plants we know so little that we cannot make any divisions of time which would be of real value in helping us to understand them.

Fossil plants from the Early Palæozoic there are, but extremely few as compared with the succeeding period, and those few but little illuminative. In the later divisions of the Pre-Carboniferous some of the plants seem to belong to the same genera as those of the Carboniferous period. There is a fern which is characteristic of one of the earlier divisions, and there are several rather indefinite impressions which may be considered as seaweeds. There is evidence also that even one of the higher groups bearing seeds (the *Cordaiteæ*) was in full swing long before the Carboniferous period began. Hence, though of Older Palæozoic plants we know little of actual fact, we can surmise the salient truths; viz., that in that period those plants must have been evolving which were important in the Devonian and Carboniferous periods; that in the earlier part of that period they did not exist, and the simpler types only clothed the earth; and that further back still, even the simpler types had not yet evolved.

Names have been given to many fragmentary bits of fossils, but for practical purposes we might as well be without them. For the present the actual plants of the Older Palæozoic must remain in a misty obscurity, their forms we can imagine, but not know.

On the other hand, of the more recent periods, those

succeeding the Carboniferous, we have a little more knowledge. Yet for all these periods, even the Tertiary immediately preceding the present day, our knowledge is far less exact and far less detailed than it is for that unique period, the Carboniferous itself.

The characteristic plants of the Carboniferous period are all very different from those of the present, and every plant of that date is now extinct. In the succeeding periods the main types of vegetation changed, and with each succeeding change advanced a step towards the stage now reached.

The Permian, geologically speaking, was a period of transition. Toward the close of the Carboniferous there were many important earth movements which raised the level of the land and tended to enclose the area of water in what is now Eastern Europe, and to make a continental area with inland seas. Many of the Carboniferous genera are found to extend through the Permian and then die out, while at the same time others became quite extinct as the physical conditions changed. The seed-bearing plants became relatively more important, and though the genus *Cordaites* died out at the end of the period it was succeeded by an increasing number of others of more advanced type.

When we come to the older MESOZOIC rocks, we have in England at any rate an area which was slowly submerging again. The more important of the plants which are preserved, and they are unfortunately all too few, are of a type which has not yet appeared in the earlier rocks, and are in some ways like the living *Cycas*, though they have many characters fundamentally different from any living type. In the vegetation of this time, plants of Cycad-like appearance seem to have largely predominated, and may certainly be taken as the characteristic feature of the period. The great Lycopod and Equisetum-like trees of the Carboniferous are represented now only by smaller individuals of the same groups, and practically all the genera which were flourishing in the Carboniferous times have become extinct.

The Cycad-like plants, however, were far more numerous and varied in character and widely spread than they ever were in any succeeding time. Still, no flowers (as we understand the word to-day) had appeared, or at least we have no indication in any fossil hitherto discovered, that true flowers were evolved until towards the end of the period (see, however, Chapter X).

The newer Mesozoic or UPPER CRETACEOUS period represents a relatively deep sea area over England, and the rocks then formed are now known as the chalk, which was all deposited under an ocean of some size whose water must have been clear, and on the whole free from ordinary débris, for the chalk is a remarkably homogeneous deposit. From the point of view of plant history, the Upper Mesozoic is notable, because in it the flowering plants take a suddenly important position. Beds of this age (though of very different physical nature) are known all over the world, and in them impressions of leaves and fruits, or their casts, are well represented. The leaves are those of both Monocotyledons and Dicotyledons, and the genera are usually directly comparable with those now living, and sometimes so similar that they appear to belong to the same genus. The cone-bearing groups of the Gymnosperms are still present and are represented by a number of forms, but they are far fewer in varieties than are the groups of flowering plants—while the Cycad-like plants, so important in the Lower Mesozoic, have relatively few representatives. There is, it almost seems, a sudden jump from the flowerless type of vegetation of the Lower Mesozoic, to a flora in the Upper Mesozoic which is strikingly like that of the present day.

The TERTIARY period is a short one (geologically speaking, and compared with those going before it), and during it the land level rose again gradually, suffering many great series of earth movements which built most of the mountain chains in Europe which are standing to the present day. In the many plant-containing deposits of this age, we find specimens indicating that the flora was very similar to the plants now living, and that flowering plants held the dominant position in the forests, as they do to-day. In fact, from the point of view of plant evolution, it is almost an arbitrary and unnecessary distinction to separate the Tertiary epoch from the present, because the main features of the vegetation are so similar. There are, however, such important differences in the distribution of the plants of the Tertiary and those of the present times, that the distinction is advisable; but it must always be remembered that it is not comparable with the wide differences between the other epochs.

Among the plants now living we find representatives of most, though not of all, of the great *groups* of plants which have flourished in the past, though in the course of time all the species have altered and those of the earliest earth periods have become extinct. The relative importance of the different groups changes greatly in the various periods, and as we proceed through the ages of time we see the dominant place in the plant world held successively by increasingly advanced types, while the plants which dominated earlier epochs dwindle and take a subordinate position. For example, the great trees of the Carboniferous period belonged to the Lycopod family, which to-day are represented by small herbs creeping along the ground. The Cycad-like plants of the Mesozoic, which grew in such luxuriance and in such variety, are now restricted to a small number of types scattered over the world in isolated localities.

During all the periods of which we have any knowledge there existed a rich and luxuriant vegetation composed of trees, large ferns, and small herbs of various kinds, but the members of this vegetation have changed fundamentally with the changing earth, and unlike the earth in her rock-forming they have never repeated themselves.

CHAPTER V

STAGES IN PLANT EVOLUTION

To attempt any discussion of the *causes* of evolution is far beyond the scope of the present work. At present we must accept life as we find it, endowed with an endless capacity for change and a continuous impulse to advance. We can but study in some degree the *course* taken by its changes.

From the most primitive beginnings of the earliest periods, enormous advance had been made before we have any detailed records of the forms. Yet there remain in the world of to-day numerous places where the types with the simplest structure can still flourish, and successfully compete with higher forms. Many places which, from the point of view of the higher plants, are undesirable, are well suited to the lower. Such places, for example, as the sea, and on land the small nooks and crannies where water drops collect, which are useless for the higher plants, suffice for the minute forms. In some cases the lower plants may grow in such masses together as to capture a district and keep the higher plants from it. Equisetum (the horsetail) does this by means of an extensive system of underground rhizomes which give the plant a very strong hold on a piece of land which favours it, so that the flowering plants may be quite kept from growing there.

In such places, by a variety of means, plants are now flourishing on the earth which represent practically all the main stages of development of plant life as a whole. It is to the study of the simpler of the living forms that we owe most of our conceptions of the course taken by evolution. Had we to depend on fossil evidence alone, we should be in almost complete ignorance of the earliest types of vegetation and all the simpler cohorts of plants, because their minute size and very delicate structure have always rendered them unsuitable for preservation in stone. At the same time, had we none of the knowledge of the numerous fossil forms which we now possess, there would be great gaps in the series which no study of living forms could supply. It is only by a study and comparison of both living and fossil plants of all kinds and from beds of all ages that we can get any true conception of the whole scheme of plant life.

Grouping together all the main families of plants at present known to us to exist or to have existed, we get the following series:—

	Group.	Common examples of typical families in the group.
Thallophyta	Algæ	Seaweeds.
	Fungi	Moulds and toadstools.
Bryophyta	Hepaticæ	Liverworts.
	Musci	Mosses.
Pteridophyta	Equisetales	Horsetails.
	Sphenophyllales *	fossil only, *Sphenophyllum*.
	Lycopodales	Club-moss.
	Filicales	Bracken fern.
Pteridospermæ	Lyginodendræ *	fossil only, *Sphenopteris*.
Gymnosperms	Cycadales	Cycads.
	Bennettitales *	fossil only, *Bennettites*.
	Ginkgoales	Maidenhair Tree.
	Cordaitales *	fossil only, *Cordaites*.
	Coniferales	Pine, Yew.
	Gnetales	Welwitschia.
Angiosperms	Monocotyledons	Lily, Palm, Grass.
	Dicotyledons	Rose, Oak, Daisy.

In this table the different groups have not a strictly equivalent scientific value, but each of those in the second column represents a large and well-defined series of primary importance, whose members could not possibly be included along with any of the other groups.

Those marked with an asterisk are known only as

STAGES IN PLANT EVOLUTION

fossils, and it will be seen that of the seventeen groups, so many as four are known only in the fossil state. This indicates, however, but a part of their importance, for in nearly every other group are many families or genera which are only known as fossils, though there are living representatives of the group as a whole.

In this table the individual families are not mentioned, because for the present we need only the main outline of classification to illustrate the principal facts about the course of evolution. As the table is given, the simplest families come first, the succeeding ones gradually increasing in complexity till the last group represents the most advanced type with which we are acquainted, and the one which is the dominant group of the present day.

This must not be taken as a suggestion that the members of this series have evolved directly one from the other in the order in which they stand in the table. That is indeed far from the case, and the relations between the groups are highly complex.

It must be remarked here that it is often difficult, even impossible, to decide which are the most highly evolved members of any group of plants. Each individual of the higher families is a very complicated organism consisting of many parts, each of which has evolved more or less independently of the others in response to some special quality of the surroundings. For instance, one plant may require, and therefore evolve, a very complex and well-developed water-carriage system while retaining a simple type of flower; another may grow where the water problem does not trouble it, but where it needs to develop special methods for getting its ovules pollinated; and so on, in infinite variety. As a result of this, in almost all plants we have some organs highly evolved and specialized, and others still in a primitive or relatively primitive condition. It is only possible to determine the relative positions of plants on the scale of development by making an average conclusion from the study of the details of all their parts. This, however, is beset with difficulties, and in most cases the scientist, weighed by personal inclinations, arbitrarily decides on one or other character to which he pays much attention as a criterion, while another scientist tends to lay stress on different characters which may point in another direction.

In no group is this better illustrated than among the Coniferæ, where the relative arrangement of the different families included in it is still very uncertain, and where the observations of different workers, each dealing mainly with different characters in the plants, tend to contradict each other.

This, however, as a byword. Notwithstanding these difficulties, which it would be unfair to ignore, the main scheme of evolution stands out clearly before the scientist of to-day, and his views are largely supported by many important facts from both fossil and living plants.

Very strong evidence points to the conclusion that the most primitive plants of early time were, like the simplest plants of to-day, water dwellers. Whether in fresh water or the sea is an undecided point, though opinion seems to incline in general to the view that the sea was the first home of plant life. It can, however, be equally well, and perhaps even more successfully argued, that the freshwater lakes and streams were the homes of the first families from which the higher plants have gradually been evolved.

For this there is no direct evidence in the rocks, for the minute forms of the single soft cells assumed by the most primitive types were just such as one could not expect to be successfully fossilized. Hence the earliest stages must be deduced from a comparative study of the simplest plants now living. Fortunately there is much material for this in the numerous waters of the earth, where swarms of minute types in many stages of complexity are to be found.

The simplest type of plants now living, which ap-

STAGES IN PLANT EVOLUTION

pears to be capable of evolution on lines which might have led to the higher plants, is that found in various members of the group of the Protococcoideæ among the Algæ. The claim of bacteria and other primitive organisms of various kinds to the absolute priority of existence is one which is entirely beyond the scope of a book dealing with fossil plants. The early evolution of the simple types of the Protococcoideæ is also somewhat beyond its scope, but as they appear to lie on the most direct "line of descent" of the majority of the higher plants it cannot be entirely ignored. From the simpler groups of the green Algæ other types have specialized and advanced along various directions, but among them there seems an inherent limitation, and none but the protococcoid forms seem to indicate the possibility of really high development.

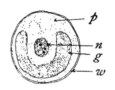

Fig. 17.—A Protococcoid Plant consisting of one cell

p, Protoplasm; *n*, nucleus; *g*, colouring body or chloroplast; *w*, cell wall.

In a few words, a typical example of one of the simple Protococcoideæ may be described as consisting of a mass of protoplasm in which lie a recognizable nucleus and a green colouring body or chloroplast, with a cell wall or skin surrounding these vital structures, a cell wall that may at times be dispensed with or unusually thickened according as the need arises. This plant is represented in fig. 17 in a somewhat diagrammatic form.

In such a case the whole plant consists of one single cell, living surrounded by the water, which supplies it with the necessary food materials, and also protects it from drying up and from immediate contact with any hard or injurious object. When these plants propagate they divide into four parts, each one similar to the original cell, which all remain together within the main cell wall for a short time before they separate.

If now we imagine that the four cells do not separate, but remain together permanently, we can see the possibility of a beginning of specialization in the different parts of the cell. The single living cell is equally acted on from all sides, and in itself it must perform all the life functions; but where four lie together, each of the four cells is no longer equally acted on from all sides. This shows clearly in the diagram of a divided cell given in fig. 18. Here it is obvious that one side of each of the four cells, viz. that named *a* in the diagram, is on the outside and in direct contact with the water and external things; but walls *b* and *c* touch only the corresponding walls of the neighbouring cells. Through walls *b* and *c* no food and water can enter directly, but at the same time they are protected from injury and external stimulus. Hence, in this group of four cells there is a slight differentiation of the sides of the cells. If now we imagine that each of the four cells, still remaining in contact, divides once more into four members, each of which reaches mature size while all remain together, then we have a group of sixteen cells, some of which will be entirely inside, and some of which will have walls exposed to the environment.

Fig. 18.— Diagram of Protococcoid Cell divided into four daughter cells. Walls *a* are external, and walls *b* and *c* in contact with each other.

If the cells of the group all divide again, in the manner shown the mass will become more than one cell thick, and the inner cells will be more completely differentiated, for they will be entirely cut off from the outside and all direct contact with water and food materials, and will depend on what the outer cells transmit to them. The outer cells will become specialized for protection and also for the absorption of the water and salts and air for the whole mass. From such a plastic group of green cells it is probable that the higher and increasingly complex forms of plants have evolved. There are still living plants which correspond with the groups of four, sixteen, &c., cells just now theoretically stipulated.

STAGES IN PLANT EVOLUTION

The higher plants of to-day all consist of very large numbers of cells forming tissues of different kinds, each of which is specialized more or less, some very

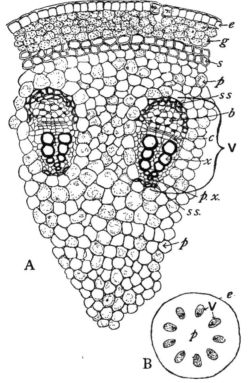

Fig. 19.—A, Details of Part of the Tissues in a Stem of a Flowering Plant. B, Diagram of the Whole Arrangement of Cross Section of a Stem: *e*, Outer protecting skin; *g*, green cells; *s*, thick-walled strengthening cells; *p*, general ground tissue cells. v, Groups of special conducting tissues: *x*, vessels for water carriage; *px*, first formed of the water vessels; *c*, growing cells to add to the tissues; *b*, food-conducting cells; *ss*, strengthening cells.

elaborately, for the performance of certain functions of importance for the plant body as a whole. With the increase in the number of cells forming the solid plant body, the number of those living wholly cut off from the

ANCIENT PLANTS

outside becomes increasingly great in comparison with those forming the external layer. Some idea of the complexity and differentiation of this cell mass is given in fig. 19, A, which shows the relative sizes and shapes of the cells composing a small part of the stem of a common flowering plant. The complete section would be circular and the groups v would be repeated round it symmetrically, and the whole would be enclosed by

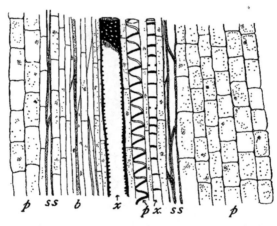

Fig. 20.—Conducting Cells and Surrounding Tissue seen in fig. 19, A, cut lengthways. *p x*, First formed vessels for water conduction; *x*, larger vessel; *b*, food-conducting cells; *s s*, strengthening cells; *p*, general ground tissue.

an unbroken layer of the cells marked *e*, as in the diagram B.

In the tissues of the higher plants the most important feature is the complex system of conducting tissues, shown in the young condition in v in fig. 19, A. In them the food and water conducting elements are very much elongated and highly specialized cells, which run between the others much like a system of pipes in the brickwork of a house. These cells are shown cut longitudinally in fig. 20, where they are lettered to correspond with the cells in fig. 19, A, with which they should be

STAGES IN PLANT EVOLUTION

compared. In such a view the great difference between the highly specialized cells x, px, b, &c., and those of the main mass of ground tissue p becomes apparent.

Even in the comparatively simply organized groups of the Equisetales and Lycopodiales the differentiation of tissues is complete. In the mosses, and still more in the liverworts, it is rudimentary; but they grow in very damp situations, where the conduction of water and the protection from too much drying is not a difficult problem for them. As plants grow higher into the air, or inhabit drier situations, the need of specialization of tissues becomes increasingly great, for they are increasingly liable to be dried, and therefore need a better flow of water and a more perfect protective coat.

It is needless to point out how the individual cells of a plant, such as that figured in figs. 19 and 20, have specialized away from the simple type of the protococcoid cell in their mature form. In the young growing parts of a plant, however, they are essentially like protococcoid cells of squarish outline, fitting closely to each other to make a solid mass, from which the individual types will differentiate later and take on the form suitable for the special part they have to play in the economy of the whole plant.

To trace the specialization not only of the tissues but of the various parts of the whole plant which have become elaborate organs, such as leaves, stems, and flowers, is a task quite beyond the present work to attempt. From the illustrations given of tissue structure from plants at the two ends of the series much can be imagined of the inevitable intermediate stages in tissue evolution.

As regards the elaboration of organs, and particularly of the reproductive organs, details will be found throughout the book. In judging of the place of any plant in the scale of evolution it is to the reproductive organs that we look for the principal criteria, for the reproductive organs tend to be influenced less by their physical surroundings than the vegetative organs, and are therefore truer guides to natural relationships.

In the essential cells of the reproductive organs, viz. the egg cell and the male cell, we get the most primitively organized cells in the plant body. In the simpler families both male and female cells return to the condition of a free-swimming protococcoid cell, and in all but the highest families the male cell requires a liquid environment, in which it *swims* to the egg cell. In the higher families the necessary water is provided within the structure of the seed, and the male cell does not swim, a naked, solitary cell, out into the wide world, as it does in all the families up to and including the Filicales. In the Coniferæ and Angiosperms the male cell does not swim, but is passive (or largely so), and is brought to the egg cell. One might almost say that the whole evolution of the complex structures found in fruiting cones and flowers is a result of the need of protection of the delicate, simple reproductive cells and the embryonic tissues resulting from their fusion. The lower plants scatter these delicate cells broadcast in enormous numbers, the higher plants protect each single egg cell by an elaborate series of tissues, and actually bring the male cell to it without ever allowing either of them to be exposed.

It must be assumed that the reader possesses a general acquaintance with the living families tabulated on p. 44; those of the fossil groups will be given in some detail in succeeding chapters which deal with the histories of the various families. It is premature to attempt any general discussion of the evolution of the various groups till all have been studied, so that this will be reserved for the concluding chapters.

CHAPTER VI

MINUTE STRUCTURE OF FOSSIL PLANTS—LIKENESSES TO LIVING ONES

The individual plants of the Coal Measure period differed entirely from those now living; they were more than merely distinct species, for in the main even the families were largely different from the present ones. Nevertheless, when we come to examine the minute anatomy of the fossils, and the cells of which they are composed, we find that between the living and the fossil cell types the closest similarity exists.

From the earliest times of which we have any knowledge the elements of the plant body have been the same, though the types of structures which they built have varied in plan. Individual *cells* of nearly every type from the Coal Measure period can be identically matched with those of to-day. In the way the walls thickened, in the shapes of the wood, strengthening or epidermal cells, in the form of the various tissues adapted to specific purposes, there is a unity of organization which it is reasonable to suppose depends on the fundamental qualities inherent in plant life.

This will be illustrated best, perhaps, by tabulating the chief modifications of cells which are found in plant tissues. The illustrations of these types in the following table are taken from living plants, because from them figures of more diagrammatic clearness can be made, and the salient characters of the cells more easily recognized. Comparison of these typical cells with those illustrated from the fossil plants reveals their identity in essential structure, and most of them will be found in the photos of fossils in these pages, though they are better recognized in the actual fossils themselves.

54   ANCIENT PLANTS

Principal Types of Plant Cells

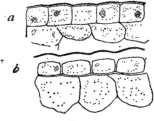

Fig. 21

EPIDERMAL.

Epidermis. — Protecting layer or skin. Cells with outer wall thickened in many cases (fig. 21, *a* and *b*). Compare fossil epidermis in fig. 34, *e*.

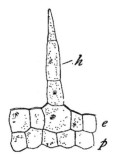

Fig. 22

Hairs. — Extensions of epidermis cells. Single cells, or complex, as fig. 22, *h*, where *e* is epidermis and *p* parenchyma. Compare fossil hairs in figs. 79 and 120.

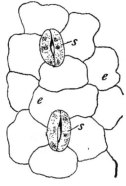

Fig. 23

Stomates. — Breathing pores in the epidermis. Seen in surface view as two-lipped structures (fig. 23). *s*, Stomates; *e*, epidermis cells. Compare fossil stomates in fig. 8.

STRUCTURE OF FOSSIL PLANTS

GROUND TISSUE

Parenchyma.—Simple soft cells, either closely packed, as in fig. 24, or with air spaces between them. Compare 78, B, for fossil.

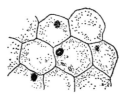

Fig. 24

Palisade. — Elongated, closely packed cells, *p*, chiefly in leaves, lying below the epidermis, *e*, fig. 25. Compare fig. 34, *p*, for fossil palisade.

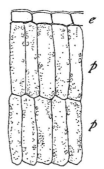

Fig. 25

Endodermis. — Cells with specially thickened walls, *en*, lying as sheath between the parenchyma, *c*, of ground tissue, and the vascular tissue, *s*, fig. 26. Compare fig. 108 for fossil endodermis.

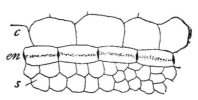

Fig. 26

Latex cells.—Large, often much elongated cells, *m*, lying in the parenchyma, *p*, fig. 27, which are packed with contents. Compare fig. 107, *s*.

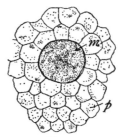

Fig. 27

56 ANCIENT PLANTS

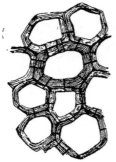

Fig. 28

Sclerenchyma. — Thick-walled cells among parenchyma for strengthening, fig. 28. Compare fig. 34, *s*.

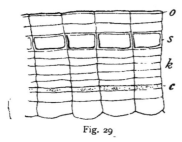

Fig. 29

Cork. — Layers of cells replacing the epidermis in old stems. Outer cells, *o*, crushed; *k*, closely packed cork cells; stone cells, *s*, fig. 29. Compare fig. 95, *k*.

Cork cambium. — Narrow, actively dividing cells, *c* in fig. 29, giving rise to new cork cells in consecutive rows.

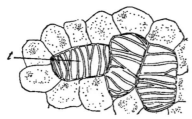

Fig. 30

Tracheides. — Specially thickened cells in the parenchyma, usually for water storage, *t*, fig. 30. Compare fig. 95, *t*.

STRUCTURE OF FOSSIL PLANTS

Vascular Tissue

Wood.—*Protoxylem*, tracheids and vessels, long, narrow elements, with spiral or ring-like thickenings, s^1 and s^2, fig. 31. Compare fig. 81, A, px, for fossil.

Metaxylem, long elements, tracheids and vessels. Some with narrow pits, as t in fig. 31; others with various kinds of pits. In transverse section seen in fig. 33, w, fossil in fig. 78, w.

Wood parenchyma.—Soft cells associated with the wood, p in fig. 31. Fossil in fig. 81, B, p.

Wood sclerenchyma.—Hard thickened cells in the wood.

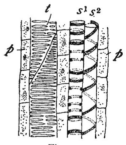

Fig. 31

Bast.—*Sieve tubes*, long cells which carry foodstuffs, cross walls pitted like sieves, s, fig. 32. In transverse section in fig. 33.

Companion cells, narrow cells with rich proteid contents, c, fig. 32. In transverse section at c, fig. 33.

Bast parenchyma.—Soft unspecialized cells mixed with the sieve tubes, p, fig. 32.

Bast fibres.—Thick-walled sclerenchymatous cells mixed with, or outside, the soft bast.

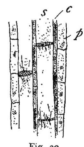

Fig. 32

Cambium.—Narrow cells, like those of the cork cambium, which lie between the wood and bast, and give rise to new tissues of each kind, cb, fig. 33. Compare fig. 114, fossil.

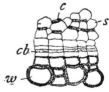

Fig. 33

There are, of course, many minor varieties of cells, but these illustrate all the main types.

Among the early fossils, however, one type of wood cell and one type of bast cell, so far as we know, are not present. These cells are the true *vessels* of the wood of flowering plants, and the long bast cells with their companion proteid cells. The figure of a metaxylem wood cell, shown in fig. 31, t, shows the more primitive type of wood cell, which has an oblique cross wall. This type of wood cell is found in all the fossil trees, and all the living plants except the flowering plants. The vessel type, which is that in the big wood vessels of the flowering plants, and has no cross wall, is seen in fig. 20, x.

The similarity between the living cells and those of the Coal Measure fossils is sufficiently illustrated to need no further comment. This similarity is an extremely helpful point when we come to an interpretation of the fossils. In living plants we can study the physiology of the various kinds of cells, and can deduce from experiment exactly the part they play in the economy of the whole plant. From a study of the tissues in any plant structure we know what function it performed, and can very often estimate the nature of the surrounding conditions under which the plant was growing. To take a single example, the palisade tissue, illustrated in fig. 25, p, in living plants always contains green colouring matter, and lies just below the epidermis, usually of leaves, but sometimes also of green stems. These cells do most of the starch manufacture for the plant, and are found best developed when exposed to a good light. In very shady places the leaves seldom have this type of cell. Now, when cells just like these are found in fossils (as is illustrated in fig. 34), we can assume all the physiological facts mentioned above, and rest assured that that leaf was growing under normal conditions of light and was actively engaged in starch-building when it was alive. From the physiological standpoint the fossil leaf is entirely the same as a normal living one.

STRUCTURE OF FOSSIL PLANTS 59

From the morphological standpoint, also, the features of the plant body from the Coal Measure period fall into the same divisions as those of the present. Roots, stems, leaves, and reproductive organs, the essentially distinct parts of a plant, are to be found in a form entirely recognizable, or sufficiently like that now in vogue to be interpreted without great difficulty. In the detailed structure of the reproductive organs more changes have

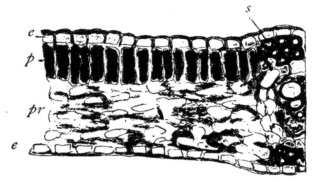

Fig. 34.—From a Photo of a Fossil Leaf

e, Epidermis; *p*, palisade cells; *pr*, soft parenchyma cells (poorly preserved); *s*, sclerenchyma above the vascular bundle.

taken place than in any others, both in internal organization and external appearance.

Already, in the Early Palæozoic period, the distinction between leaves, stems, roots, and reproductive organs was as clearly marked as it is to-day, and, judging by their structure, they must each have performed the physiological functions they now do. Roots have changed least in the course of time, probably because, in the earth, they live under comparatively uniform conditions in whatever period of the world's history they are growing. Naturally, between the roots of different species there are slight differences; but the likeness between fern roots from the Palæozoic and from a living fern is absolutely complete. This is illustrated in fig. 35, which shows the microscopic structure of the two roots when cut in transverse direction. The various tissues will be recognized as coming into the table on p. 54, so that both in the details of individual cells and in the general arrangement of the cell groups or tissues the roots of these fossil and living ferns agree.

Among stems there has been at all periods more variety than among the roots of the corresponding plants, and in the following chapter, when the differences between living and fossil plants will be considered, there

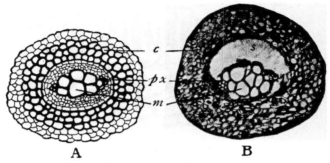

Fig. 35.—A, Root of Living Fern. B, Root of Palæozoic Fossil Fern. *c*, Cortex; *p x*, protoxylem in two groups; *m*, metaxylem; *s*, space in fossil due to decay of soft cells.

will be several important structures to notice. Nevertheless, there are very many characters in which the stems from such widely different epochs agree. The plants in the palæozoic forests were of many kinds, and among them were those with weak trailing stems which climbed over and supported themselves on other plants, and also tall, sturdy shafts of woody trees, many of which were covered with a corky bark. Leaves were attached to the stems, either directly, as in the case of some living plants, or by leaf stalks. In external appearance and in general function the stems then were as stems are now. In the details of the individual cells also the likeness is complete; it is in the grouping of the cells, the anatomy of the tissues, that the important differences lie. It has

STRUCTURE OF FOSSIL PLANTS

been remarked already that increase in complexity of the plant form usually goes with an increase in complexity of the cells and variety of the tissues. The general ground tissue in nearly all plants is very similar; it is principally in the vascular system that the advance and variety lie.

Plant anatomists lay particular stress on the vascular system, which, in comparison with animal anatomy, holds an even more important position than does the skeleton. To understand the essential characters of stems, both living and fossil, and to appreciate their points of likeness or difference, it is necessary to have some knowledge of the general facts of anatomy; hence the main points on which stress is laid will be given now in brief outline.

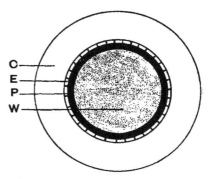

Fig. 36.—Diagram of Simplest Arrangement of Complete Stele in a Stem

W, Central solid wood; P, ring of bast; E, enclosing sheath of endodermis; C, ground tissue or cortex.

Leaving aside consideration of the more rudimentary and less defined structure of the algæ and mosses, all plants may be said to possess a "vascular system". This is typically composed of elongated wood (or xylem) with accessory cells (see p. 57, table), and bast (phloem), also with accessory cells. These specialized conducting elements lie in the ground tissue, and in nearly all cases are cut off from direct contact with it by a definite sheath, called the endodermis (see p. 55, fig. 26). Very often there are also groups or rings of hard thick-walled cells associated with the vascular tissues, which protect them and play an important part in the consolidation of the whole stem.

The simplest, and probably evolutionally the most

primitive form which is taken by the vascular tissues, is that of a single central strand, with the wood in the middle, the bast round it, and a circular endodermis enclosing all, as in fig. 36, which shows a diagram of this arrangement. Such a mass of wood and bast surrounded by an endodermis, is technically known as a *stele*, a very convenient term which is much used by anatomists. In its simplest form (as in fig. 36) it is called a *protostele*, and is to be found in both living and fossil plants. A number of plants which get more complex steles later on, have protosteles in the early stages of their development, as in *Pteris aurita* for example, a species allied to the bracken fern, which has a hollow ring stele when mature.

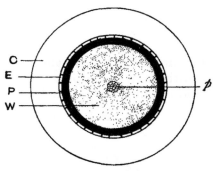

Fig. 37.—Diagram of a Stele with a few Cells of Pith p in the Middle of the Wood. Lettering as in fig. 36

The next type of stele is quite similar to the protostele, but with the addition of a few large unspecialized cells in the middle of the wood (p, fig. 37); these are the commencement of the hollowing process which goes on in the wood, resulting later in the formation of a considerable pith, as is seen in fig. 38, where the wood is now a hollow cylinder, as the phloem has been from the first. When this is the case, a second

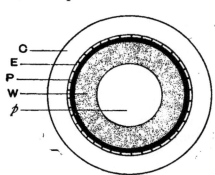

Fig. 38.—Diagram showing Extensive Pith p in the Wood. Lettering as in fig. 36

sheath or endodermis generally develops on the inner side of the wood, outside the pith, and cuts the vascular tissues off from the inner parenchyma. A further step is the development of an inner cylinder of bast so that the vascular ring is completely double, with endodermis on both sides of the cylinder, as is seen in fig. 39.

In all these cases there is but one strand or cylinder, of vascular tissue in the stem, but one stele, and this type of anatomy is known as the *monostelic* or single-steled type.

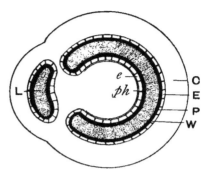

Fig. 39.—A Cylindrical Stele, with e, inner endodermis, and *ph*, inner phloem; W, wood; P, outer phloem; E, outer endodermis. L, part of the stele going out to supply a large leaf, thus breaking what would otherwise appear as a closed ring stele

When from the double cylinder just described a strand of tissue goes off to supply a large leaf, a considerable part of the stele goes out and breaks the ring. This is shown in fig. 39, where L is the part of the stele going to a leaf, and the rest the broken central cylinder. When the stem is short, and leaves grow thickly so that bundles are constantly going out from the main cylinder, this gets permanently broken, and its appearance when cut across at any given point is that of a group of several steles arranged in a ring, each separate stele being like the simple protostele in its structure. See fig. 40. This type of stem has long been known as *polystelic* (*i.e.* many-steled), and

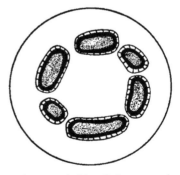

Fig. 40.—A Ring Stele apparently broken up into a Number of Protosteles by many Leaf Gaps

it is still a convenient term to describe it by. There has been much theoretical discussion about the true meaning of such a "polystelic" stem, which cannot be entered into here; it may be noted, however, that the various strands of the broken ring join up and form a meshwork when we consider the stem as a whole, it is only in a single section that they appear as quite independent protosteles. Nevertheless, as we generally consider the anatomy of stems in terms of single sections, and as the descriptive word "polystelic" is a very convenient and widely understood term, it will be used throughout the book when speaking of this type of stem anatomy.

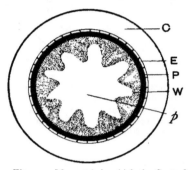

Fig. 41.—Monostele in which the Central Pith is Star-shaped, and the Wood breaking up into Separate Groups

p, Pith; w, wood; P, phloem; E, endodermis; C, cortex.

Such a type as this, shown in fig. 40, is already complex, but it often happens that the steles branch and divide still further, until there is a highly complicated and sometimes bewildering system of vascular strands running through the ground tissue in many directions, but cut off from it by their protective endodermal sheaths. Such complex systems are to be found both in living and fossil plants, more especially in many of the larger ferns (see fig. 88).

Higher plants in general, however, and in particular flowering plants, do not have a polystelic vascular arrangement, but a specialized type of monostele.

Referring again to fig. 37 as a starting-point, imagine the pith in the centre to spread in a star-shaped form till the points of the star touched the edges of the ring, and thus to break the wood ring into groups. A stage in this process (which is not yet completed) is shown in fig. 41, while in fig. 42 the wood and bast groups

STRUCTURE OF FOSSIL PLANTS

are entirely distinct. In the flowering plants the cells of the endodermis are frequently poorly characterized, and the pith cells resemble those of the cortical ground tissue, so that the separate groups of wood and bast (usually known as "vascular bundles", in distinction from the "steles" of fig. 40) appear to lie independently in the ground tissue. These strands, however, must not be confused with steles, they are only fragments of the single apparently broken up stele which runs in the stem.

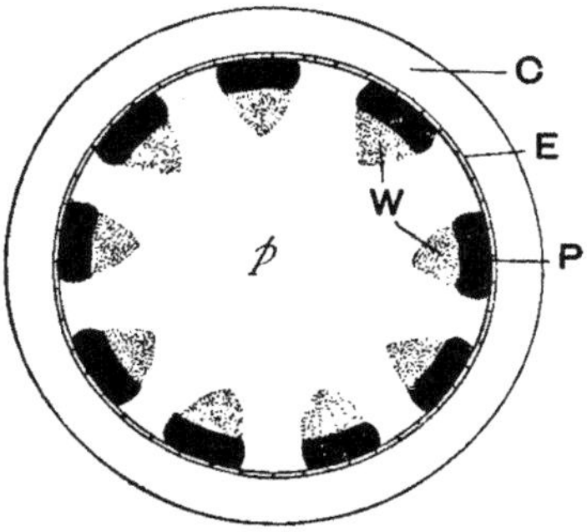

Fig. 42.—Monostele in which the Pith has invaded all the Tissues as far as the Endodermis, and broken the Wood and Phloem up into Separate Bundles. These are usually called "vascular bundles" in the flowering plants

The vascular bundle, of all except the Monocotyledons, has a potentiality for continued growth and expansion which places it far above the stele in value for a plant of long life and considerable growth. The cells lying between the wood and the bast, the soft parenchyma cells always accompanying such tissues, retain their vitality and continue to divide with great regularity, and to give rise to a continuous succession of new cells of wood on the one side and bast on the other; see fig. 33, c, b. In this way the primary, distinct vascular bundles are joined by a ring of wood, see fig. 43, to which are added further rings every season, till the mass of wood becomes a strong solid shaft.

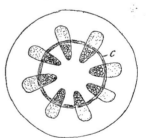

Fig. 43.—Showing actively growing Zone c (Cambium) in the Vascular Bundles, and joining across the ground tissue between them

This ever-recurring activity of the cambium gives rise to what are known as "annual rings" in stems, see fig. 44, in which the wood shows both primary distinct groups in the centre, and the rings of growth of later years.

Cambium with this power of long-continued activity is found in nearly all the higher plants of to-day (except the Monocotyledons), but in the fern and lycopod groups it is in abeyance. Certain cases from nearly every family of the Pteridophytes are known, where some slight development of cambium with its secondary thickening takes place, but in the groups below the Gymnosperms cambium has almost no part to play. On the other hand, so far back as the Carboniferous period, the masses of wood in the Pteridophyte trees were formed by cambium in just the same way as they are now in the higher forms. Its presence was almost universal at that time in the lower groups where to-day there are hardly any traces of it to be found.

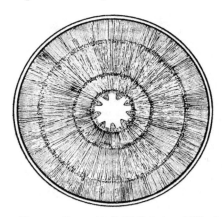

Fig. 44.—Stem with Solid Cylinder of Wood developed from the Cambium, showing three "annual rings" In the centre may still be seen the separate groups of the wood of the primary "vascular bundles"

It will be seen from this short outline of the vascular system of plants, that there is much variety possible from modifications of the fundamental protostele. It is also to be noted that the plants of the Coal Measures had already evolved all the main varieties of steles which are known to us even now,[1] and that the development of secondary thickening was very widespread. In several cases the complexity of type exceeds that of

---

[1] Though the Angiosperm was not then evolved, the Gymnosperm stem has distinct vascular bundles arranged as are those of the Angiosperm, the difference here lies in the type of wood cells.

STRUCTURE OF FOSSIL PLANTS

modern plants (see Chap. VII), and there are to be found vascular arrangements no longer extant.

When we turn to the *Reproductive Organs*, we find that the points of likeness between the living and the fossil forms are not so numerous or so direct as they are in the case of the vegetative system.

Fig. 45.—Fern Sporangia

A, fossil; B, living.

As has been indicated, the families of plants typical of the Coal Measures were not those which are the most prominent to-day, but belonged to the lower series of Pteridophytes. In their simpler forms the fructifications then and now resemble each other very closely, but in the more elaborate developments the points of variety are more striking, so that they will be dealt with in the following chapter. Cases of likeness are seen in the sporangia of ferns, some of which appear to have been practically identical with those now living. This is illustrated in fig. 45, which shows the outline of the cells of the sporangia of living and fossil side by side.

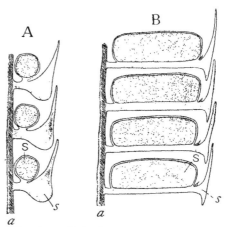

Fig. 46.—A, Living Lycopod cone; B, *Lepidodendron* (fossil) cone. *a*, Axis; *s*, scale; S, sporangium with spores. One side of a longitudinal section

In the general structure also of the cones of the simpler types of *Lepidodendron* (fossil, see frontispiece) there is a close agreement with the living Lycopods, though as regards size and output of spores there was a considerable difference in favour of the fossils. The plan of each is that round the axis of the cone simple scales are arranged, on each of which, on its upper side, is seated a large sporangium bearing numerous spores all of one kind (see fig. 46).

Equally similar are the cones of the living Equisetum and some of the simple members of the fossil family Calamiteæ, but the more interesting cases are those where differences of an important morphological nature are to be seen.

As regards the second [1] generation there is some very important evidence, from extremely young stages, which has recently been given to the world. In a fern sporangium *germinating spores* were fossilized so as to show the first divisions of the spore cell. These seem to be identical with the first divisions of some recent ferns (see fig. 47). This is not only of interest as showing the close similarity in detail between plants of such widely different ages, but is a remarkable case of delicate preservation of soft and most perishable structures in the "coal balls".

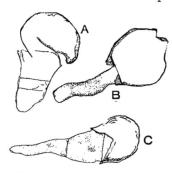

Fig. 47.—Germinating Fern Spores

A and B, from carboniferous fossils; C, living fern. (A and B after Scott.)

While these few cases illustrate points of likeness between the fructifications of the Coal Measures and of to-day, the large size and successful character of the primitive Coal Measure plants was accompanied by many developments on the part of their reproductive organs which are no longer seen in living forms, and the greater

---

[1] The gametophyte generation (represented in the ferns by the prothallium on which the sexual organs develop) alternates with the large, leafy sporophyte. Refer to Scott's volume on *Flowerless Plants* (see Appendix) for an account of this alternation of generations.

number of palæozoic fructifications must be considered in the next chapter.

CHAPTER VII

MINUTE STRUCTURE OF FOSSIL PLANTS—DIFFERENCES FROM LIVING ONES

We have seen in the last chapter that the main morphological divisions, roots, stems, leaves, and fructifications, were as distinct in the Coal Measure period as they are now. There is one structure, however, found in the Coal Measure fossils, which is hardly paralleled by anything similar in the living plants, and that is the fossil known as *Stigmaria*. *Stigmaria* is the name given, not to a distinct species of plant, but to the large rootlike organs which we know to have belonged to all the species of *Lepidodendron* and of *Sigillaria*. In the frontispiece these organs are well seen, and branch away at the foot of the trunk, spreading horizontally, to all appearance merely large roots. They are especially regularly developed, however, the main trunk giving rise always to four primary branches, these each dividing into two equal branches, and so on—in this they are unlike the usual roots of trees. They bore numerous rootlets, of which we know the structure very well, as they are the commonest of all fossils, but in their internal anatomy the main "roots" had not the structure which is characteristic of roots, but were like *stems*. In living plants there are many examples of stems which run underground, but they always have at least the rudiments of leaves in the form of scales, while the fossil structures have apparently no trace of even the smallest scales, but bear only rootlets, thus resembling true roots. The questions of morphology these structures raise are too complex to be discussed here, and

Stigmaria is only introduced as an example, one of the very few available, of a palæozoic structure which seems to be of a nature not clearly determinable as either root, stem, leaf, or fructification. Among living plants the fine rootlike rhizophores of Selaginella bear some resemblance to Stigmaria in essentials, though so widely different from them in many ways, and they are probably the closest analogy to be found among the plants of to-day.

Fig. 48.—Stele of *Lepidodendron* w, surrounded by a small ring of secondary wood s

The individual cells, we have already seen, are strikingly similar in the case of fossil and living plants. There are, of course, specific varieties peculiar to the fossils, of which perhaps the most striking seem to be some forms of *hair* cells. For example, in a species of fern from the French rocks there were multicellular hairs which looked like little stems of Equisetum owing to regular bands of teeth at the junctions of the cells. These hairs were quite characteristic of the species—but hairs of all sorts have always abounded in variety, so that such distinction has but minor significance.

As was noted in the table (p. 58) the only cell types of prime importance which were not evolved by the Palæozoic plants were the wood vessels, phloem and accompanying cells which are characteristic of the flowering plants.

Among the fossils the vascular arrangements are most interesting, and, as well as all the types of stele development noted in the previous chapter as common to both living and fossil plants, there are further varieties found only among the fossils (see fig. 50).

The simple protostele described (on p. 61) is still found, particularly in the very young stages of living

STRUCTURE OF FOSSIL PLANTS

ferns, but it is a type of vascular arrangement which is not common in the mature plants of the present day. In the Coal Measure period, however, the protostele was characteristic of one of the two main groups of ferns. In different species of these ferns, the protostele assumed a large variety of shapes and forms as well as the simple

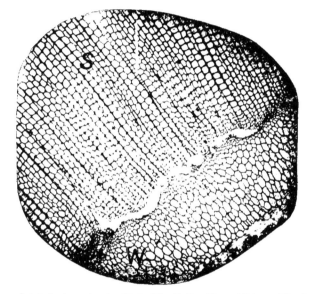

Fig. 49.—*Lepidodendron*, showing Part of the Hollow Ring of Primary Wood W, with a relatively large amount of Secondary Tissue S, surrounding it

cylindrical type. The central mass of wood became five-rayed in some, star-shaped, and even very deeply lobed, with slightly irregular arms, but in all these cases it remained fundamentally monostelic. Frequently secondary tissue developed round the protosteles of plants whose living relatives have no such tissue. A case of this kind is illustrated in fig. 48, which shows a simple circular stele surrounded by a zone of secondary woody tissue in a species of *Lepidodendron*.

In many species of *Lepidodendron* the quantity of

secondary wood formed round the primary stele was very great, so that (as is the case in higher plants) the primary wood became relatively insignificant compared with it. In most species of *Lepidodendron* the primary stele is a hollow ring of wood (cf. fig. 38, p. 62) round which the secondary wood developed, as is seen in fig. 49. These two cases illustrate a peculiarity of fossil plants. Among living ones the solid and the simple ring stele are almost confined to the Pteridophytes, where secondary wood does not develop, but the palæozoic Pteridophytes, while having the simple primary types of steles, had quantities of secondary tissue, which was correlated with their large size and dominant position.

Among *polystelic* types (see p. 63) we find interesting examples in the fossil group of the *Medulloseæ*, which are much more complex than any known at present, both owing to their primary structure and also to the peculiar fact that all the steles developed secondary tissue towards the inner as well as the outer side. One of the simpler members of this family found in the English Coal Measures is illustrated in fig. 50. Here there are three principal protosteles (and several irregular minor ones) each of which has a considerable quantity of secondary tissue all round it, so that a portion of the secondary wood is growing in towards the actual centre of the stem as a whole—a very anomalous state of affairs.

In the more complex Continental type of *Medullosa* there are *very* large numbers of steles. In the one figured from the Continent in fig. 51 but a few are represented. There is a large outer double-ring stele, with

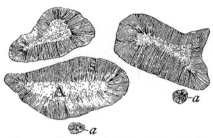

Fig. 50.—Diagram of Steles of the English *Medullosa*, showing three irregular, solid, steles A, with secondary thickenings s, all round each. *a*, Small accessory steles

STRUCTURE OF FOSSIL PLANTS

secondary wood on both sides of it, and within these a number of small steles, all scattered through the ground tissue, and each surrounded by secondary wood. In actual specimens the number of these central steles is much greater than that indicated in the diagram.

No plant exists to-day which has such an arrangement of its vascular cylinder. It almost appears as though at the early period, when the Medulloseæ flourished, steles were experimenting in various directions. Such types as are illustrated in figs. 50 and 51 are obviously wasteful (for secondary wood developing towards the centre of a stem is bound to finally meet), and complex, but apparently inefficient, which may partly account for the fact that this type of structure has not survived to the present, though simpler and equally ancient types have done so.

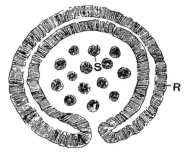

Fig. 51.—Continental *Medullosa*, showing R, outer double-ring stele with secondary wood all round it; S, inner stellate steles, also surrounded in each case by secondary tissue

Further details of the anatomy of fossils will be mentioned when we come to consider the individual families; those now illustrated suffice to show that in the Coal Measures very different arrangements of steles were to be found, as well as those which were similar to those existing now. The significance of these differences will become apparent when their relation to the other characters of the plants is considered.

The fructifications, always the most important parts of the plant, offer a wide field, and the divergence between the commoner palæozoic and recent types seems at first to be very great. Indeed, when palæozoic reproductive bodies have to be described, it is often necessary to use the common descriptive terms in an altered and wider sense.

Among the plants of to-day there are many varieties of the simple single-celled reproductive masses which are called *spores*, and which are usually formed in large numbers inside a spore case or sporangium. Among the higher plants *seeds* are also known in endless variety, all of which, compared with spores, are very complex, for they are many-celled structures, consisting essentially of an embryo or young plant enclosed in various protective coats. The distinction between the two is sharp and well defined, and for the student of living plants there exists no difficulty in separating and describing seeds and spores.

But when we look back through the past eras to palæozoic plants the subject is not so easy, and the two main types of potentially reproductive masses are not sharply distinct. The seed, as we know it among recent plants, and as it is generally defined, had not fully evolved; while the spores were of great variety and had evolved in several directions, some of which seem to have been intermediate stages between simple spores and true seeds. These seedlike spores served to reproduce the plants of the period, but their type has since died out and left but two main methods among living plants, namely the essentially simple spores, the very simplicity of whose organization gives them a secure position, and the complex seeds with their infinite variety of methods for protecting and scattering the young embryos they contain.

Among the Coal Measure fossils we can pick up some of the early stages in the evolution of the seed from the spore, or at least we can examine intermediate stages between them which give some idea of the possible course of events. Hence, though the differences from our modern reproductive structures are so noticeable a feature of the palæozoic ones, it will be seen that they are really such differences as exist between the members at the two ends of a series, not such as exist between unrelated objects.

STRUCTURE OF FOSSIL PLANTS

Very few types can be mentioned here, and to make their relations clear a short series of diagrams with explanations will be found more helpful than a detailed account of the structures.

Fig. 52.—Spores

Each spore a single cell which develops with three others in tetrads (groups of four). Very numerous tetrads enclosed in a spore case or sporangium which develops on a leaf-like segment called the sporophyll. Each spore germinates independently of the others after being scattered, all being of the same size. Common in fossils and living Pteridophytes.

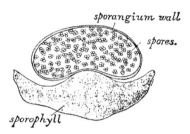

Fig. 53.—Spores

Each a single cell like the preceding, but here only one tetrad in a sporangium ripens, so that each contains only four spores. Compared with the preceding types these spores are very large. Otherwise details similar to above. Some fossils have such sporangia with eight spores, or some other small number; living Selaginellas have four. In the same cone sporangia with small spores are developed and give rise to the male organs.

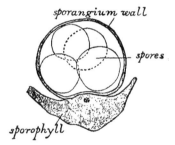

Fig. 54.—"Spores" of Seedlike Structure

Out of a tetrad in each sporangium only one spore ripens, s in figure, the others, s, abort. The wall of the sporangium, w, is more massive than in the preceding cases, and from the sporophyll, flaps, spf, grow up on each side and enclose and protect the sporangium. The one big spore appears to germinate inside these protective coats, and not to be scattered separately from them. Only found in fossils, one of the methods of reproduction in *Lepidodendron*. Other sporangia with small spores were developed which gave rise to the male organs.

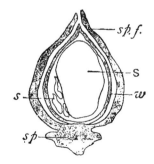

76   ANCIENT PLANTS

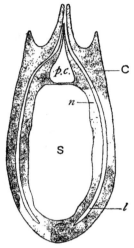

Fig. 55.—"Seed"

In appearance this is like a seed, but differs from a true seed in having no embryo, and is like the preceding structure in having a very large spore, s, though there is no trace of the three aborting ones. The spore develops in a special mass of tissue known as the nucellus, *n*, which partly corresponds to the sporangium wall of the previous types. In it a cavity, *p c*, the pollen chamber, receives the pollen grains which enter at the apex of the "seed." There is a complex coat, C, which stands round the nucellus but is not joined to it, leaving the space *l* between them. Only in fossils; *Trigonocarpus* (see p. 122) is similarly organized. Small spores in fern-like sporangia, called pollen grains.

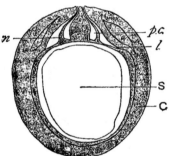

Fig. 56.—"Seed"

Very similarly organized to the above, but the coat is joined to the nucellus about two-thirds of its extent, and up to the level *l*. In the pollen chamber, *p c*, a cone of nucellar tissue projects, and the upper part of the coat is fluted, but these complexities are not of primary importance. The large spore s germinated and was fertilized within the "seed", but apparently produced no embryo before it ripened. Small "spores" in fern-like sporangia form the pollen grains. Only in fossils, *e.g.* Lagenostoma. (See p. 119.)

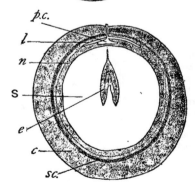

Fig. 57.—Seed

Essentially similar to the preceding, except in the possession of an embryo *e*, which is, however, small in comparison with the endosperm which fills the spore s. The whole organization is simpler than in the fossil *Lagenostoma*, but the coat is fused to the nucellus further up (see *l*). Small "spores" form the pollen grains. Living and fossil type, Cycads and Ginkgo.

STRUCTURE OF FOSSIL PLANTS

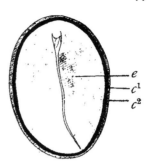

Fig. 58.—Seed

In the ripe seed the large embryo e practically fills up all the space within the two seed coats c^1 and c^2; endosperm, pollen chamber, &c., have been eliminated, and the young ovule is very simple and small as a result of the protection and active service of the carpels in which it is enclosed. Small "spores" form the pollen grains. Typical of living Dicotyledons.

These few illustrations represent only the main divisions of an army of structures with an almost unimaginable wealth of variety which must be left out of consideration.

For the structures illustrated in figs. 54, 55, and 56 we have no name, for their possible existence was not conceived of when our terminology was invented, and no one has yet christened them anew with distinct names. They are evidently too complex in organization and too similar to seeds in several ways to be called spores, yet they lack the essential element in a seed, namely, an embryo. The term "ovule" (usually given to the young seed which has not yet developed an embryo) does not fit them any better, for their tissues are ripened and hard, and they were of large size and apparently fully grown and mature.

For the present a name is not essential; the one thing that is important is to recognize their intermediate character and the light they throw on the possible evolution of modern seeds.

A further point of great interest is the manner in which these "seeds" were borne on the plant. To-day seeds are always developed (with the exception of Cycas) in cones or flowers, or at least special inflorescences. But the "seed" of *Lagenostoma* (fig. 56), as well as a number of others in the group it represents, were not borne on a special structure, but directly on the green foliage leaves. They were in this on a level with the simple sporangia of ferns which appear on the backs of the fronds, a fact which is of great significance both for our views on the evolution of seeds as such, and for the bearing it has on the relationships of the various groups of allied plants. This will be referred to subsequently (Chapter XI), and is mentioned now only as an example of the difference between some of the characters of early fossils and those of the present day.

It is true that botanists have long recognized the organ which bears seeds as a modified leaf. The carpels of all the higher plants are looked on as *homologous* with leaves, although they do not appear to be like them externally. Sometimes among living plants curious diseases cause the carpels to become foliar, and when this happens the diseased carpel reverts more or less to the supposed ancestral leaf-like condition. It is only among the ancient (but recently discovered) fossils, however, that seeds are known to be borne normally on foliage leaves.

From Mesozoic plants we shall learn new conceptions about flowers and reproductive inflorescences in general, but these must be deferred to the consideration of the family as a whole (Chapter XIII).

Enough has been illustrated to show that though the individual cells, the bricks, so to speak, of plant construction, were so similar in the past and present, yet the organs built up by them have been continually varying, as a child builds increasingly ambitious palaces with the same set of bricks.

CHAPTER VIII
PAST HISTORIES OF PLANT FAMILIES

1. Flowering Plants, Angiosperms

In comparison with the other groups of plants the flowering families are of recent origin, yet in the sense in which the word is usually used they are ancient indeed, and the earliest records of them must date at least to periods hundreds of thousands of years ago.

Through all the Tertiary period (see p. 34) there were numerous flowering plants, and there is evidence that many families of both Monocotyledons and Dicotyledons existed in the Upper Cretaceous times. Further back than this we have little reliable testimony, for the few specimens of so-called flowering plants from the Lower Mesozoic are for the most part of a doubtful nature.

The flowering plants seem to stand much isolated from the rest of the plant world; there is no *direct* evidence of connection between their oldest representatives and any of the more primitive families. So far as our actual knowledge goes, they might have sprung into being at the middle of the Mesozoic period quite independently of the other plants then living; though there are not wanting elaborate and almost convincing theories of their connection with more than one group of their predecessors (see p. 108).

It is a peculiarly unfortunate fact that although the rocks of the Cretaceous and Tertiary are so much less ancient than those of the Coal Measures, they have preserved for us far less well the plants which were living when they were formed. Hitherto no one has found in Mesozoic strata masses of exquisitely mineralized Angiosperm fragments[1] like those found in the Coal Measures,

---

[1] Material recently obtained by the author and Dr. Fujii in Japan does contain some true petrifactions of Angiosperms and other plant debris. The account of these discoveries has not yet been published.

which tell us so much about the more ancient plants. Cases are known of more or less isolated fragments with their microscopical tissues mineralized. For example, there are some palms and ferns from South America which show their anatomical structure very clearly preserved in silica, and which seem to resemble closely the living species of their genera. The bulk of the plants preserved from these periods are found in the form of casts or impressions (see p. 10), which, as has been pointed out already, are much less satisfactory to deal with, and give much less reliable results than specimens which have also their internal structure petrified. The quantity of material, however, is great, and impressions of single leaves innumerable, and of specimens of leaves attached to stems, and even of flowers and fruits, are to be found in the later beds of rock. These are generally clearly recognizable as belonging to one or other of the living families of flowering plants. Leaf impressions are by far the most frequent, and our knowledge of the Tertiary flora is principally derived from a study of them. Their outline and their veins are generally preserved, often also their petioles and some indication of the thickness and character of the fleshy part of the leaf. From the outline and veins alone an expert is generally able to determine the species to which the plant belongs, though it is not always quite safe to trust to these determinations or to draw wide-reaching conclusions from them.

In fig. 59 is shown a photograph of the impression of a Tertiary leaf, which illustrates the condition of an average good specimen from rocks of the period. Its shape and the character of the veins are sufficient to mark it out immediately as belonging to the Dicotyledonous group of the flowering plants.

Seeds and fruits are also to be found; and in some very finely preserved specimens from Japan stamens from a flower and delicate seeds are seen clearly impressed on the light stone. In fig. 60 is illustrated a

PAST HISTORIES OF PLANT FAMILIES

Fig. 59.—Dicotyledonous Leaf Impression from Tertiary Rocks

couple of such seeds, which show not only their wings but also the small antennæ-like stigmas. Specimens so perfectly preserved are practically as good as herbarium material of recent plants, and in this way the externals of the Tertiary plants are pretty well known to us.

Fig. 60.—Seeds from Japanese Tertiary Rocks; at a are seen the two stigmas still preserved

A problem which has long been discussed, and which has aroused much interest, is the relative antiquity of the Monocotyledonous and the Dicotyledonous branches of the flowering plants. A peculiar fascination seems to hang over this still unsolved riddle, and a battle of flowers may be said to rage between the lily and the rose for priority. Recent work has thrown no decisive light on the question, but it has undoubtedly demolished the old view which supposed that the Monocotyledons (the lily group) appeared at a far earlier date upon this earth than the Dicotyledons. The old writers based their contention on incorrectly determined fossils. For instance, seeds from the Palæozoic rocks were described as Monocotyledons because of the three or six ribs which were so characteristic of their shell; we know now that these seeds (*Trigonocarpus*) belong to a family already mentioned in another connection (p. 72), the Medulloseæ (see p. 122), the affinity of which lies between the cycads and the ferns. Leaves of *Cordaites*, again, which are broad and long with well-marked parallel veins, were described as those of a Monocotyledonous plant like the Yucca of to-day; but we now know them to belong to a family of true Gymnosperms possibly distantly related to *Taxus* (the Yew tree).

Recent work, which has carefully sifted the fossil evidence, can only say that no true Monocotyledons have yet been found below the Lower Cretaceous rocks, and that at that period we see also the sudden inrush of Dicotyledons. Hence, so far as palæontology can show, the two parallel groups of the flowering plants arose about the same time. It is of interest to note, however, that the only petrifaction of a flower known from any part of the world is an ovary which seems to be that of one of the Liliaceæ. In the same nodules, however, there are several specimens of Dicotyledonous woods, so that it does not throw any light on the question of priority.

With the evidence derived from the comparative study of the anatomy of recent flowering plants we cannot concern ourselves here, beyond noting that the results weigh in favour of the Dicotyledons as being the more primitive, though not necessarily developed much earlier in point of time. Until very much more is discovered than is yet known of the origin of the

PAST HISTORIES OF PLANT FAMILIES 83

flowering plants as a whole, it is impossible to come to a more definite conclusion about this much-discussed subject.

Let us now attempt to picture the vegetable communities since the appearance of the flowering plants. The facts which form the bases of the following conceptions have been gathered from many lands by numerous workers in the field of fossil botany, from scattered plant remains such as have been described.

When the flowering plants were heralded in they appeared in large numbers, and already by the Cretaceous period there were very many different species. Of these a number seem to belong to genera which are still living, and many of them are extremely like living species. It would be wearisome and of little value to give a list of all the recorded species from this period, but a few of the commoner ones may be mentioned to illustrate the nature of the plants then flourishing.

Several species of *Quercus* (the Oak) appeared early, particularly *Quercus Ilex*; leaves of the *Juglandaceæ* (Walnut family) were very common, and among the Tertiary fossils appear its fruits. Both *Populus* (the Poplar) and *Salix* (the Willow) date from the early rocks, while *Ficus* (the Fig) was very common, and *Casuarina* (the Switch Plant) seems to have been widely spread. Magnolias also were common, and it appears that *Platanus* (the Plane) and *Eucalyptus* coexisted with them.

It will be immediately recognized that the above plants have all living representatives, either wild or cultivated, growing in this country at the present day, so that they are more or less familiar objects, and there appears to have been no striking difference between the early flowering plants and those of the present day. Between the ancient Lycopods, for example, and those now living the differences are very noteworthy; but the earliest of the known flowering plants seem to have been essentially like those now flourishing. It must be remembered in this connection that the existing flowering plants are immensely nearer in point of time to their origin than are the existing Lycopods, and that when such æons have passed as divide the present from the Palæozoic, the flowering plants of the future may have dwindled to a subordinate position corresponding to that held by the Lycopods now.

A noticeable character of the early flowering-plant flora, when taken as a whole, is the relatively large proportion of plants in it which belong to the family *Amentiferæ* (oaks, willows, poplars, &c.). This is supposed by some to indicate that the family is one of the most primitive stocks of the Angiosperms. This view, however, hardly bears very close scrutiny, because it derives its main support from the large numbers of the Amentiferæ as compared with other groups. Now, the Amentiferæ were (and are) largely woody resistant plants, whose very nature would render them more liable to be preserved as impressions than delicate trees or herbs, which would more readily decay and leave no trace. Similarly based on uncertain evidence is the surmise that the group of flowers classed as *Gamopetalæ* (flowers with petals joined up in a tube, like convolvulus) did not flourish in early times, but are the higher and later development of the flower type. Now, *Viburnum* (allied to the honeysuckle) belongs to this group, and it is found right down in the Cretaceous, and *Sambucus* (Elder, of the same family) is known in the early Tertiary. These two plants are woody shrubs or small trees, while many others of the family are herbs, and it is noteworthy that it is just these woody, resistant forms which are preserved as fossils; their presence demonstrates the antiquity of the group as a whole, and the absence of other members of it may be reasonably attributed to accidents of preservation. In the Tertiary also we get a member of the heath family, viz. *Andromeda*, and another tube-flower, *Bignonia*, as well as several more *woody* gamopetalous flowers.

PAST HISTORIES OF PLANT FAMILIES 85

Hence it is wise to be very cautious about drawing any important conclusions from the relative numbers of the different species, or the absence of any type of plant from the lists of those as yet known from the Cretaceous. When quantities of structurally preserved material can be examined containing the flowering plants in petrifactions, then it will be possible to speak with some security of the nature of the Mesozoic flora as a whole.

The positive evidence which is already accumulated, however, is of great value, and from it certain deductions may be safely made. Specimens of Cretaceous plants from various parts of the world seem to indicate that there was a very striking uniformity in the flora of that period all over the globe. In America and in Central Europe, for example, the same types of plants were growing. We shall see that, as time advanced, the various types became separated out, dying away in different places, until each great continent and division of land had a special set of plants of its own. At the commencement of the reign of flowering plants, however, they seem to have lived together in the way we are told the beasts first lived in the garden of Eden.

At the beginning of the Tertiary period there were still many tropical forms, such as Palms, Cycads, *Nipa*, various *Artocarpaceæ, Lauraceæ, Araliaceæ*, and others, growing side by side with such temperate forms as *Quercus, Alnus, Betula, Populus, Viburnum*, and others of the same kind. Before the middle of the Tertiary was reached the last Cycads died in what is now known as Europe; and soon after the middle Tertiary all the tropical types died out of this zone.

At the same time those plants whose leaves appear to have fallen at the end of the warm season began to become common, which is taken as an indication of a climatic influence at work. Some writers consider that in the Cretaceous times there was no cold season, and therefore no regular period of leaf fall, but as the climate became temperate the deciduous trees increased in numbers; yet the Gymnospermic and Angiospermic woods which are found with petrified structure show well-marked annual rings and seem to contradict this view.

Toward the end of the Tertiary times there were practically no more tropical forms in the European flora, though there still remained a number of plants which are now found either only in America or only in Asia.

The Glacial epoch at the close of the Tertiary appears to have driven all the plants before it, and afterwards, when its glaciers retreated, shrinking up to the North and up the sides of the high mountains, the plant species that returned to take possession of the land in the Quaternary or present period were those which are still inhabiting it, and the floras of the tropics, Asia, and America were no longer mixed with that of Europe.[1]

CHAPTER IX

PAST HISTORIES OF PLANT FAMILIES

II. Higher Gymnosperms

The more recent history of the higher Gymnosperms, in the Upper Cretaceous and Tertiary periods, much resembles that of the flowering plants as sketched in the previous chapter. Many of the genera appear to have been those still living, and some of the species even may have come very close to or have been identical with those of to-day. The forms now characteristic of the different continents were growing together, and appear to have been widely distributed over the globe. For example, *Sequoia* and *Taxodium*, two types now characteristic of America, and *Glyptostrobus*, at present found

---

[1] A fuller account of the Angiospermic flora can be had in French, in M. Laurent's paper in *Progressus Rei Botanicæ*. See Appendix for reference.

in Asia, were still growing with the other European types in Europe so late as middle Tertiary times.

As in the case of the Angiosperms, the fossils we have of Cretaceous and Tertiary Gymnosperms are nearly all impressions and casts, though some more or less isolated stems have their structure preserved. Hence our knowledge of these later Gymnosperms is far from complete. From the older rocks, however, we have both impressions and microscopically preserved material, and are more fully acquainted with them than with those which lived nearer our own time. Hard, resistant leaves, which are so characteristic of most of the living genera of Gymnosperms, seem to have been also developed in the past members of the group, and these tend to leave clear impressions in the rocks, so that we have reliable data for reconstructing the external appearance of the fossil forms from the Palæozoic period.

The resinous character of Gymnosperm wood probably greatly assisted its preservation, and fragments of it are very common in rocks of all ages, generally preserved in silica so as to show microscopic structure. The isolated wood of Gymnosperms, however, is not very instructive, for from the wood alone (and usually it is just fragments of the secondary wood which are preserved) but little of either physiological or evolutional value can be learned. When twigs with primary tissues and bark and leaves attached are preserved, then the specimens are of importance, for their true character can be recognized. Fortunately among the coal balls there are many such fragments, some of which are accompanied by fruits and male cones, so that we know much of the Palæozoic Gymnosperms, and find that in some respects they differ widely from those now living.

There is, therefore, much more to be said about the fossil Gymnosperms than about the Angiosperms, both because of the better quality of their preservation and because their history dates back to a very much earlier period than does the Angiospermic record. Indeed, we do not know when the Gymnosperms began; the well-developed and ancient group of *Cordaiteæ* was flourishing before the Carboniferous period, and must therefore date back to the rocks of which we have no reliable information from this point of view, and the origin of the Gymnosperms must lie in the pre-Carboniferous period.

The group of Gymnosperms includes a number of genera of different types, most of which may be arranged under seven principal families. In a sketch of this nature it is, of course, quite impossible to deal with all the less-important families and genera. Those that will be considered here are the following:—

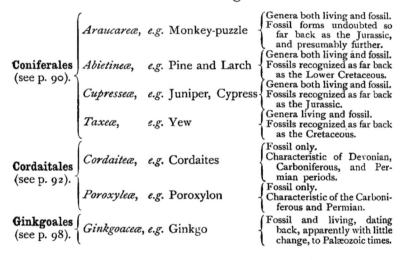

We must pay the most attention to the two last groups, as they are so important as fossils, and the *Cordaiteæ* were a very numerous family in Coal Measure times. They had their period of principal development so long ago that it is probable that no direct descendants remain to the present time, though some botanists consider that the *Taxeæ* are allied to them.

Of the groups still living it is difficult, almost impos-

PAST HISTORIES OF PLANT FAMILIES 89

sible, to say which is the highest, the most evolved type. In the consideration of the Gymnosperm family it is brought home with great emphasis how incomplete and partial our knowledge is as yet. Many hold that the *Araucareæ* are the most primitive of the higher Gymnosperms. In support of this view the following facts are noted. They have a simple type of fructification, with a single seed on a simple scale, and many scales arranged round an axis to form a cone. In the microscopic structure of their wood they have double rows of bordered pits, a kind of wood cell which comes closer to the old fossil types than does the wood of any of the other living genera. Further than this, wood which is almost indistinguishable from the wood of recent Araucarias is found very far back in the rocks, while their leaves are broad and simple, and attached directly to the stem in a way similar to the leaves of the fossil *Cordaiteæ*, and very different from the needle leaves on the secondary stems of the Pine family; so that there appears good ground for considering the group an ancient and probably a primitive one.[1]

On the other hand, there are not wanting scientists who consider the *Abietineæ* the living representatives of the most primitive and ancient stock, though on the whole the evidence seems to indicate more clearly that the Pine-tree group is specialized and highly modified. Their double series of foliage leaves, their complex cones (whose structures are not yet fully understood), and their wood all support the latter view.

Some, again, consider the *Taxeæ* as a very primitive group, and would place them near the Cordaiteæ, with which they may be related. Their fleshy seeds, growing not in cones but on short special axes, support this view, and it is certainly true that in many ways the large seeds, with their succulent coats and big endosperm, are much

---

[1] From the Cretaceous deposits of North America several fossil forms (*Brachyphyllum*, *Protodammara*) are described which show clear affinities with the family as it is now constituted. (See Hollick and Jeffrey; reference in the Appendix.)

like those of the lower Gymnosperms and of several fossil types. Those, however, who hold to the view that the Abietineæ are primitive, see in the *Taxeæ* the latest and most modified type of Gymnosperm.

It will be seen from this that there is no lack of variety regarding the interpretation of Gymnosperm structures.

The Gymnosperms do not stand in such an isolated position as do the Angiosperms. Whatever the variety of views held about the details of the relative placing of the families within the group, all agree in recognizing the evidence which enables us to trace with confidence the connection between the lower Gymnosperms and the families of ferns. There are many indications of the intimate connection between higher and lower Gymnosperms. Between the series exist what might be described as different degrees of cousinship, and in the lower groups lie unmistakable clues to their connection with more ancient groups in the past which bridge over the gaps between them and the ferns.

For the present, however, let us confine ourselves to the history of the more important Gymnosperms, the discussion of their origin and the groups from which they may have arisen must be postponed until the necessary details about those groups have been mentioned.

To a consideration of the living families of *Araucareæ*, *Abietineæ*, *Cupresseæ*, and *Taxeæ* we can allow but a short space; their general characters and appearance are likely to be known to the reader, and their details can be studied from living specimens if they are not. For purposes of comparison with the fossils, however, it will be necessary to mention a few of the principal features which are of special importance in discussing phylogeny.

The ARAUCARIACEÆ are woody trees which attain a considerable size, with broad-based, large leaves attached directly to the stem. In the leaves are a series of numerous parallel vascular bundles. The wood cells in microscopic section show two rows or more of round bordered pits. The cones are very large, but the male and female are different in size and organization. The female cone is composed of series of simple scales arranged spirally round the axis, and each scale bears a single seed and a small ligule.

The pollen grains from the male cone are caught on the ligule and the pollen tubes enter the micropyle of the ovule, bringing in passive male cells which may develop in large numbers in each grain. The seeds when ripe are stony, and some are provided with a wing from part of the tissue of the scale. In the ripe cones the scales separate from the cone axis.

The ABIETINEÆ are woody trees, some reaching a great height, all with a strong main stem. The leaves are of two kinds: primary ones borne directly attached to the stem (as in first-year shoots of the Larch), and secondary ones borne in tufts of two (in Pine) or a large number (in older branches of Larch) on special short branches, the primary leaves only developing as brown scales closely attached to the stems. Leaves generally very fine and needlelike, and with a central vascular bundle. The wood in microscopic section shows a single row of round bordered pits on the narrow tracheæ.

The female cones are large, male and female differing greatly in size and organization. The female cone, composed of a spiral series of pairs of scales, which often fuse together as the cone ripens. Each upper scale of the pair bears two seeds. The pollen grains from the male cone enter the micropyle of the seed and are caught in the tissue (apex of nucellus) there; the pollen tubes discharge passive male cells, only two of which develop in each grain. The seeds when ripe are stony and provided with a wing from the tissue of the scale on which they were borne.

The CUPRESSEÆ are woody trees reaching no great height, and of a bushy, branching growth. The leaves are attached directly to the main stem, and arrange themselves in alternating pairs of very small leaves, closely pressed to the stem. The wood in microscopic section shows a single row of round bordered pits on the tracheæ.

The cones are small, and the scales forming them arranged in cycles. The female scales bear a varying number of seeds. The pollen grain has two passive male cells. The seeds when ripe are stony, with wings, though in some cases (species of Juniper) the cone scales close up and become fleshy, so that the whole fruit resembles a berry.

The TAXEÆ are woody, though not great trees, bushily branched. The leaves are attached spirally all round the stem, but place themselves so as to appear to lie in pairs arranged in one horizontal direction. The wood in microscopic section shows a single row of round bordered pits on the tracheæ.

There are small male cones, but the seeds are not borne on cones, growing instead on special short axes, where there may be several young ovules, but on which usually two seeds ripen. The seeds are big, and have an inner stone and outer fleshy covering. Some have special outer fleshy structures known as "arils", *e.g.* the red outer cup round the yew "berry" (which is not a berry at all, but a single unenclosed seed with a fleshy coat).

When we turn to the CORDAITEÆ we come to a group of plants which bears distinct relationship to the preceding, but which has a number of individual characters. It is a group of which we should know nothing were it not for the fossils preserved in the Palæozoic rocks; yet, notwithstanding the fact that it flourished so long ago, it is a family of which we know much. At the time of the Coal Measures and the succeeding Permo-carboniferous period, it was of great importance, and, indeed, in some of the French deposits it would seem as though whole layers of coal were composed entirely of its leaves.

PAST HISTORIES OF PLANT FAMILIES

Among the fossil remains of this family there are impressions, casts, and true petrifactions, so that we know both its external appearance and the internal anatomy of nearly every part of several species of the genus. For a long time the various fossil remains of the plant were not recognized as belonging to each other and together forming the records of one and the same plant—the broad, long leaves with their parallel veins were looked on as Monocotyledons (see fig. 61); the pith casts (see fig. 63) were thought to be peculiar constricted stems, and were called *Sternbergia*; while the wood, which was known from its microscopic structure, was called *Araucarioxylon*— but the careful work of many masters of fossil botany, whose laborious studies we cannot describe in detail here, brought all these fragments together and proved them to belong to *Cordaites*.

We now know that *Cordaites* were large trees, with strong upright shafts of wood, to whose branches large simple leaves were attached. The leaves were much bigger than those of any living Gymnosperm, even than those of the Kauri Pine (a member of the Araucariaceæ), and seem in some species to have exceeded 3 ft. in length. The trees branched only at the top of the main shaft, and with their huge swordlike leaves must have differed greatly in appearance from any plant now living. The leaves had many parallel veins, as can be seen in fig. 61, and were attached by a broad base directly to the main stem; thus coming closer to the

Fig. 61.—Leaf of *Cordaites*, *l*, attached by its broad base to a Stem, *s*

Araucarias than the other groups of Gymnosperms in their leaf characters.

The internal anatomy is often well preserved, and

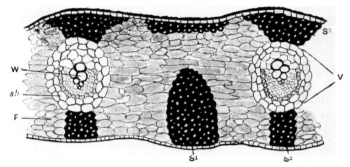

Fig. 62A.—Microscopic Section of Part of a Leaf of *Cordaites*

V, Vascular bundle; W, wood of bundle; *sh*, its sheath; S¹, large sclerenchyma mass alternating with bundles; S² and S³, sclerenchyma caps of bundle; P, soft tissue of leaf.

there is a number of species of leaves whose anatomy is known. As will be expected from the parallel veins, in each section there are many vascular bundles running equidistantly through the tissue. Fig. 62A shows the microscopic details from a well-preserved leaf. In all the species patches of sclerenchyma were developed, and everything indicates that they were tough and well protected against loss of water, even to a greater extent than are most of the leaves of living Gymnosperms.

Fig. 62B.—Much-magnified Wood Elements from *Cordaites* Stem seen in longitudinal section, the type known as *Araucarioxylon*. Note the hexagonal outlines of the bordered pits, which lie in several rows

In the stems the pith was much larger than that in living Gymnosperms (where the wood is generally very solid), and it was hollow in older stems, except for discs of tissue across the cavity. The internal cast from these stems has been described before, and is seen in fig. 63.

The wood was formed in closely packed radiating

PAST HISTORIES OF PLANT FAMILIES 95

rows by a normal cambium (see p. 66), and the tracheæ so formed had characteristic rows of bordered pits (see fig. 62B). The wood comes nearer to that of the living Araucarias than any other, and indeed the numerous pieces of fossil wood of this type which are known from all the geological periods are called *Araucarioxylon*.[1] A double strand goes out from the main mass of wood, which afterwards divides and subdivides to provide the numerous bundles of the leaf.

In the case of these fossils we are fortunate enough to have the fructifications, both male and female, in a good state of preservation. As in other Gymnosperms, the male and female cones are separate, but they differed less from each other in their arrangement than do those of any of the living types hitherto mentioned. They can hardly be described as true cones, though they had something of that nature; the seeds seem to be borne on special short stems, round which are also sterile scales. In the seed and the way it is borne perhaps the Cordaiteæ may be compared more nearly with the Taxeæ than with the other groups.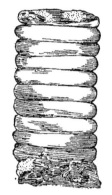

Fig. 63.—Cast of Hollow Pith of *Cordaites*, the constrictions corresponding to discs of solid tissue across the cavity

A seed, not yet ripe, is shown in slightly diagrammatic form in fig. 64, where the essential details are illustrated. The seeds of this family sometimes reached a considerable size, and had a fleshy layer which was thick in comparison with the stone, and externally comparable with a cherry—though, of course, of very different nature in reality, for *Cordaites*, like *Taxus*, is a Gymnosperm, with simple naked seeds, while a cherry is the fruit of an Angiosperm.

[1] The addition of *-oxylon* to the generic name of any living type indicates that we are dealing with a fossil which closely resembles the living type so far as we have information from the petrified material.

In a few words, these are the main characters of the large group of *Cordaites*, which held the dominant position among Gymnosperms in the Palæozoic era. They have relationships, or perhaps one should say likenesses, to many groups. Their stem- and root-anatomy is similar to the Coniferæ of the present day, the position of the ovules is like that in the Taxaceæ, the male cones in some measure recall those of *Ginkgo*, the anatomy of their leaves has points which are comparable with those of the Cycads, to which group also the large pith in the stem and the structure of some details in the seeds unite them. Their own specially distinctive characters lie in their crown of huge leaves, and unbranched shaft of stem, the similarity of their male and female inflorescences, and some points in their pollen grains which have not been mentioned. The type is a very complex one, possibly coming near the stock which, having branched out in various directions, gave rise to several of the living families.

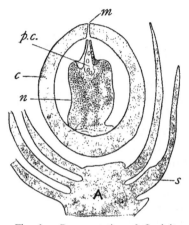

Fig. 64.—Representation of *Cordaites* Seed and its Axis with Scales, slightly diagrammatic, modified from Renault.

A, Axis with *s*, scales; *c*, coat of the seed, from which the inner parts have shrunk away; *n*, nucellus; *p.c*, pollen chamber containing pollen grains which enter through *m*.

Plants which come very near to the Cordaiteæ are the POROXYLEÆ. Of this group we have unfortunately no remains of fructifications in organic connection, so that its actual position must remain a little doubtful till they are discovered. There seems no doubt that they must have borne seeds.

Still, it has been abundantly demonstrated in recent years that the anatomy of the root, stem, and leaves

PAST HISTORIES OF PLANT FAMILIES

indicates with considerable exactness the position of any plant, so that, as these are known, we can deduce from them, with a feeling of safety, the position that *Poroxylon* takes in the natural system. In its anatomy the characters are those of the Cordaiteæ, with certain details which show a more primitive nature and seem to be characteristic of the groups below it in organization.

Poroxylon is not common, and until recently had not been found in the Lower Coal Measures of England. The plants appear to have been much smaller than *Cordaites*, with delicate stems which bore relatively large simple leaves. The anatomy of the root was that common in Gymnosperms, but the stem had a very large pith, and the leaves were much like those of *Cordaites* in having parallel veins. An important character in the anatomy of the stem was the presence of what is known as *centripetal wood*.

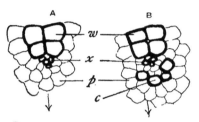

Fig. 65.—A, Normal bundle of higher plant; *x*, protoxylem on inner side next the pith *p*, and the older wood *w* outside it, *centrifugal* wood. B, Bundle with wood cells *c* developed on inner side of protoxylem, *centripetal* wood; the arrow indicates the direction of the centre of the stem.

This must be shortly explained. In all the stems hitherto considered, the first-formed wood cells (protoxylems, see p. 57) developed at the central point of the wood, towards the pith (see fig. 19, *px*, p. 49). This is characteristic of all Angiosperms and the higher Gymnosperms (except in a couple of recently investigated Pines), but among the lower plants we find that part of the later wood develops to the inner side of these protoxylem masses. The distinction is shown in fig. 65.

This point is one to which botanists have given much attention, and on which they have laid much weight in considering the affinities of the lower Gymnosperms and the intermediate groups between them and the ferns, which are found among the fossils. In *Cordaites* this point of connection with the lower types is not seen, but in *Poroxylon*, which has otherwise a stem anatomy very similar to *Cordaites*, we find groups of *centripetal* wood developed inside the protoxylem of primary bundles. For this reason, principally, is *Poroxylon* of interest at present, as in its stem anatomy it seems to connect the *Cordaites* type with that of the group below it in general organization.

GINKGOALES.—Reference to p. 44 shows that *Ginkgo*, the Maidenhair tree, belongs to the Ginkgoales, a group taking equal rank with the large and complex series of the Coniferales. The Ginkgoales of the present day, however, have but one living representative. *Ginkgo* stands alone, the single living species of its genus, representing a family so different from any other living family that it forms a prime group by itself.

Had the tree not been held sacred in China and Japan, it is probable that it would long since have been extinct, for it is now known only in cultivation. It is indeed a relic from the past which has been fortunately preserved alive for our examination. It belongs to the fossil world, as a belated November rose belongs to the summer.

Because of its beauty and interest the plant is now widely distributed under cultivation, and is available for study almost as freely as the other types of living Gymnosperms already mentioned, so that but a short summary of its more important features is needed here.

Old plants, such as can be seen growing freely in Japan (in Kew Gardens there is also a fine specimen), are very tall handsome woody trees, with noble shafts and many branches. The leaves grow on little side shoots and are the most characteristic external feature of the tree; their living form is illustrated in fig. 66, which shows the typical simple shape as well as the lobed form of the leaf which are to be found, with all intermediate stages, on the same tree. No other plant (save a

PAST HISTORIES OF PLANT FAMILIES 99

few ferns, which can generally be distinguished from it without difficulty) has leaves at all like these, so that it

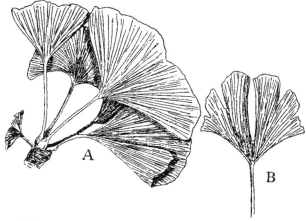

Fig. 66.—A, Tuft of *Ginkgo* Leaves, showing their "maidenhair"-like shape. B, Single deeply-divided Leaf to be found on the same tree, usually on young branches.

is particularly easy to identify the fossil remains, of which there are many.

The wood is compact and fine grained, the rings of secondary tissue being developed from a normal cambium as in the case of the higher Gymnosperms, and the individual tracheæ have round bordered pits. There are small male cones, but the seeds are not borne in cones. They develop on special stalks on which are no scales, but a small mass of tissue at the base of the seed called the "collar". Usually there are two young ovules, of which often only one ripens to a fleshy seed, though both may mature.

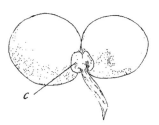

Fig. 67.—Ripe Stage of *Ginkgo* Seeds attached to their Stalk. *c*, "Collar" of seed.

The ripe seed reaches the size shown in the diagram, and is orange coloured and very fleshy; within it is a stone encasing the endosperm, which is large, *green*, and starchy, and contains the embryo with two cotyledons. This embryo is small compared with the endosperm, cf. fig. 57, p. 76, which is somewhat similar to that of *Ginkgo* in this stage.

Of the microscopic characters of the reproductive organs the most remarkable is the male cell. This is not a passive nucleus, as in the plants hitherto considered, but is an *actively swimming* cell of some size, provided with a spiral of cilia (hairlike structures) whose movements propel it through the water. In the cavity of the unripe seed these swim towards the female cell, and actively penetrate it. The arrangements of the seed are diagrammatically shown in fig. 68, which should be compared with that of *Cycas*, fig. 76, with which it has many points in common.

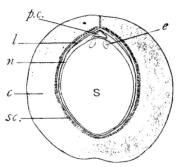

Fig. 68.—Section through Seed of *Ginkgo*

p.c, Pollen chamber in the nucellus *n*, which is fused to the coat *c* to the level *l*; *sc*, stony layer in coat; s, the big spore, filled with endosperm tissue (in this case green in colour); *e*, egg cells, one of which will produce the embryo after fertilization.

The nature of the male cell in *Cordaites* is not yet known, but there is reason to suspect it may have been actively swimming also. As this is uncertain, however, we may consider *Ginkgo* the most highly organized plant which has such a primitive feature, a feature which is a bond of union between it and the ferns, and which, when it was discovered about a dozen years ago, caused a considerable sensation in the botanical world.

To turn now to the fossil records of this family. Leaf impressions of *Ginkgo* are found in rocks of nearly all ages back even to the Upper Palæozoic. They show a considerable variety of form, and it is certain that they do not all belong to the same *species* as the living plant,

PAST HISTORIES OF PLANT FAMILIES

but probably they are closely allied. Fig. 69 shows a typical impression from the Lower Mesozoic rocks. In this specimen, the cells of the epidermis were fortunately sufficiently well preserved to be seen with the microscope, and there is a distinct difference in the size and shape of the cells of living and fossil species, see fig. 70; but this difference is slight as compared with the great similarity of form and appearance, as can be seen on comparing figs. 69 and 66, B, so that the fossil is at the most a different species of the genus *Ginkgo*. Among the fossil leaves there is greater variety than among the living ones, and some which are very deeply lobed so as to form a divided palm-like leaf go by different names, e.g. *Baiera*, but they are supposed to belong to the same family. Fossil seeds and male cones are also known as impressions, and are found far back in the Mesozoic rocks. From the fossil impressions it is certain that *Ginkgo* and plants closely allied to it were very widespread in the past, as they are found all over Europe as well as the other continents. Particularly in the Lower Mesozoic rocks *Ginkgo* seems to have been a world-wide type growing in great abundance.

In the Palæozoic the records are not so undoubted, but there is strong evidence which leads us to suppose that if the genus now living were not then extant, at least other closely related genera were, and there seems to be good grounds for

Fig. 69.—Leaf Impression of *Ginkgo* from Mesozoic Rocks of Scotland

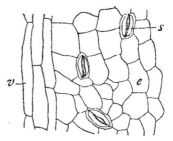

Fig. 7c.—Showing Epidermis with Stomates from the lower side of the Leaf seen in fig. 69

e, Epidermis cells; *s*, stomates; *v*, long cells of epidermis lying over the veins.

supposing that *Ginkgo* and *Cordaites* may have both arisen from some ancient common stock.

CHAPTER X

PAST HISTORIES OF PLANT FAMILIES

III. The Bennettitales

This fascinating family is known only from the fossils, and is so remote in its organization from any common living forms that it may perhaps be a little difficult for those who do not know the Cycads to appreciate the position of Bennettites. It would probably be better for one studying fossil plants for the first time to read the chapters on the Cycads, Pteridosperms, and Ferns before this chapter on the present group, which has characters connecting it with that series.

Until recently the bulk of the fossils which are found as impressions of stems and foliage of this family were very naturally classed as Cycads. They are extremely common in the Mesozoic rocks (the so-called Age of Cycads), and in the external appearance of both stems and leaves they are practically identical with the Cycads.

A few incomplete fructifications of some species have been known in Europe for many years, but it is only recently that they have been fully known. This is owing to Wieland's[1] work on the American species, which has made known the complete organization of the fructifications from a mass of rich and well-petrified material.

In the Lower Cretaceous and Upper Jurassic rocks of America these plants abound, with their microscopic structure well preserved, and their fructifications show an organization of a different nature from that of any past or present Cycad.

[1] See reference in the Appendix to this richly illustrated volume.

Probably owing to their external appearance, Wieland describes the plants as "Cycads" in the title of his big book on them; but the generic name he uses, *Cycadeoidea*, seems less known in this country than the equally well-established name of *Bennettites*, which has long been used to denote the European specimens of this family, and which will be used in the following short account of the group.

At the present time no family of fossils is exciting more interest. Their completely Cycadean appearance and their unique type of fructification have led many botanists to see in them the forerunners of the Angiosperms, to look on them as the key to that mystery—the origin of the flowering plants. This position will be discussed and the many facts in its favour noted, but we must not forget that the *Bennettitales* have only recently been realized fully by botanists, and that a new toy is ever particularly charming, a new cure particularly efficacious, and a new theory all-persuasive.

From their detailed study of the flowering plants botanists have leaned toward different groups as the present representatives of the primitive types. The various claims of the different families to this position cannot be considered here; probably that of the Ranales (the group of families round Ranunculaceæ as a central type) is the best supported. Yet these plants are most frequently delicate herbs, which would have stood relatively less chance of fossilization than the other families which may be considered primitive. They are peculiarly remote from the group of Bennettiteæ in their vegetative structure, a fact the importance of which seems to have been underrated, for in the same breath we are assured that the Bennettites are a kind of cousin to the ancient Angiosperms, and that the Ranales are among the most primitive living Angiosperms, and therefore presumably nearest the ancient ones.

However, let us leave the charms of controversy on one side and look at the actual structure of the group.

They were widely spread in Lower Mesozoic times, the plants being preserved as casts, impressions, and with structure in great numbers. The bulk of the described structural specimens have been obtained from the rocks of England, France, Italy, and America, although leaf impressions are almost universally known. The genus *Williamsonia* belongs to this family, and is one of the best known of Mesozoic plant impressions.

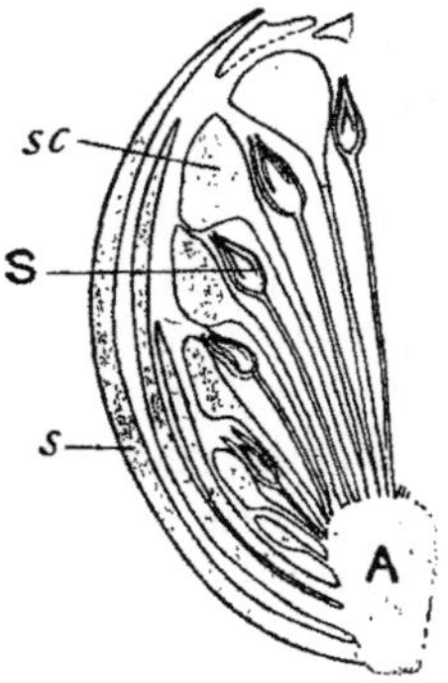

Externally the Bennettiteæ were identical in appearance with stumpy Cycads, and their leaves it is which gave rise to the surmise, so long prevalent, that the Lower Mesozoic was the "Age of Cycads", just as it was the Pteridosperm leaves that gave the Palæozoic the credit of being the "Age of Ferns". In the anatomy of both stem and leaf, also, the characters are entirely Cycadean; the outgoing leaf trace is indeed simpler in its course than that of the Cycads.

Fig. 71.—Half of a Longitudinal Section through a Mature Cone of *Bennettites*

A, Short conical axis; s, enclosing bracts; S, seeds; sc, sterile scales between the seeds.

The fructifications, however, differ fundamentally from those of the Cycads, as indeed they do from those of any known family. They took the form of compact cones, which occurred in very large numbers in the mature plants hidden by the leaf bases. In *Williamsonia*, of which we know much less detail, the fructifications stood away from the main axis on long pedicels.

In *Bennettites* the cones were composed of series of sheathing scales surrounding a short conical axis on which stood thin radiating stalks, each bearing a seed. Between them were long-stalked sterile scales with expanded ends. A part of a cone is illustrated diagrammatically in fig. 71. The whole had much the appearance of a complex fruit. In some specimens these features alone are present in the cones, but in younger

PAST HISTORIES OF PLANT FAMILIES 105

cones from the American plants further structures are found attached. Below the main axis of the seed-bearing part of the cone was a series of large complex leaf-like structures closely resembling fern leaves in their much-divided nature. On the pinnæ of these leaves were crowded innumerable large sporangia, similar to

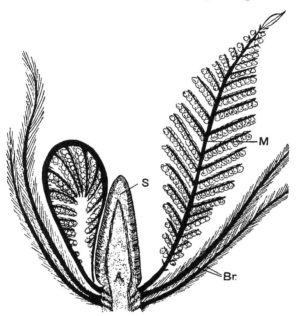

Fig. 72.—Diagram of Complete Cone of *Bennettites*

A, Central axis of conical shape terminating in the seed-bearing cone s. (After Wieland), and bearing successively Br., bracts, comparable with floral leaves; M, large complex leaves with pollen sacs.

those of a fern, which provided the pollen grains. The fossils are particularly well preserved, and have been found with these male (pollen-bearing) organs in the young unopened stages, and also in the mature unfolded condition, as well as the ripening seed cones from which they have faded, just as the stamens fade from a flower when the seeds enlarge.

It appears that these huge complex leaflike structures were really stamens, but nevertheless they were rolled up in the circinate form as are young fern leaves, and as they unrolled and spread out round the central cone they must have had the appearance of a whorl of leaves (see fig. 72).

This, in a few words, is the main general character of the fructification. The most important features, on which stress is laid, are the following. The association of the male and female structures on the same axis, with the female part *above* the male. This arrangement is found only in the flowering plants; the lower plants, which have male and female on the same cone, have them mixed, or the female below, and are in any case much simpler in their entire organization. The conical form of the axis is also important, as is the fact that it terminates in the seed-bearing structures.

Fig. 73.—Diagram of Cross Section of *Bennettites* Seed, with Embryo

c, Double-layered seed coat; *n*, crushed nucellus; *cot.*, two cotyledons which practically fill the seed.

The position of the individual seeds, each on the end of a single stalk, is remarkable, as are the long-stalked bracts whose shield-like ends join in the protection of the seeds. These structures together give the cone much of the appearance of a complex fruit of a flowering plant, but the structure of the seeds themselves is that of a simple Gymnosperm.

In the seeds, however, was an *embryo*. In this they differ from all known seeds of an earlier date, which, as has been already noted (see p. 77), are always devoid of one. This embryo is one of the most important features of the plant. It had two cotyledons which filled the seed space (see fig. 73), and left almost no trace of the endosperm. Reference to p. 112 will show that this is an advance on the Cycad seed, which has a small embryo embedded in a large mass of endosperm, and that it practically coincides with the Dicotyledonous type.

The seed with its embryo suggested comparison with the Angiosperms long before the complete structure of the fructification was known.

The fern-like nature of the pollen-bearing structures is another very important point. Were any one of these leaflike "stamens" found isolated its fern-like nature would not have been questioned a year or two ago, and their presence in the "flower" of *Bennettites* is a strong argument in favour of the Fern-Pteridosperm affinities of the group.

Had the parts of this remarkable fructification developed on separate trees, or on separate branches or distinct cones of the same one, they would have been much less suggestive than they are at present, and the fructifications might well have been included among those of the Gymnosperms, differing little more (apart from the embryo) from the other Gymnosperm genera than they do from each other. In fact, the extremely fern-like nature of the male organs is almost more suggestive of a Pteridosperm affinity, for even the simplest Cycads have well-marked scaly cones as their male organs. The female cone, again, considered as an isolated structure, can be interpreted as being not vitally different from *Cordaites*, where the seeds are borne on special short stalks amidst scales.

The embryo would, in any case, point to a position among advanced types; but it is so common for one organ of a plant to evolve along lines of its own independently, or in advance of the other organs, that the embryo structure alone could not have been held to counterbalance the Cycadean stems and leaves, the Pteridosperm-like male organs, and the Gymnospermic seeds.

But all these parts occur on the same axis, arranged in the manner typical of Angiosperms. The seed-bearing structures at the apex, the "stamens" below them, and a series of expanded scales below these again, which it takes little imagination to picture as incipient petals and sepals; and behold—the thing is a flower!

And being a "flower", is in closest connection with the ancestors of the modern flowering plants, which must consequently have evolved from some Cycadean-like ancestor which also gave rise to the Bennettitales. Thus can the flowering plants be linked on to the series that runs through the Cycads directly to the primitive ferns!

It is evident that this group, of all those known among the fossils, comes most closely to an approximation of Angiospermic structure and arrangement. Enough has been said to show that in their actual nature they are not Angiosperms, though they have some of their characters, while at the same time they are not Cycads, though they have their appearance. They stand somewhere between the two. Though many botanists at present hold that this mixture of characters indicates a relationship equivalent to a kind of cousinship with the Angiosperms, and both groups may be supposed to have originated from a Cycadean stock, this theory has not yet stood the test of time, nor is it supported by other evidence from the fossils. We will go so far as to say that it appears as though *some* Angiosperms arose in that way; but flowering plants show so many points utterly differing from the whole Cycadean stock that a little scepticism may not be unwholesome.

It is well to remember the Lycopods, where (as we shall see, p. 141) structures very like seeds were developed at the time when the Lycopods were the dominant plants, and we do not find any evidence to prove that they led on to the main line of seed plants. Similarly, Cycads may have got what practically amounted to flowers at the time when they were the dominant group, and it is very conceivable that they did not lead on to the main line of flowering plants.

Whatever view may be held, however, and whatever may be the future discoveries relating to this group of plants, we can see in the Bennettitales points which throw much light on the potentialities of the Cycadean stock, and structures which have given rise to some most interesting speculations on the subject of the Angiosperms. This group is another of the jewels in the crown of fossil botany, for the whole of its structures have been reconstructed from the stones that hold all that remains of this once extensive and now extinct family of plants.

CHAPTER XI

PAST HISTORIES OF PLANT FAMILIES

IV. The Cycads

The group of the *Cycadales*, which has a systematic value equivalent to the *Ginkgoales*, contains a much larger variety of genera and species than does the latter. There are still living nine genera, with more than a hundred and fifty species, which form (though a small one compared with most of the prime groups) a well-defined family. They are the most primitive Gymnosperms, the most primitive seed-bearing plants now living, and in their appearance and characters are very different from any other modern type. Their external resemblance to the group of the Bennettitales, however, is very striking, and indeed, without the fructifications it would be impossible to distinguish them.

The best known of the genera is that of *Cycas*, of which an illustration is given in fig. 74. The thick, stumpy stem and crown of "palm"-like leaves give it a very different appearance from any other Gymnosperm. Commonly the plants reach only a few feet in height, but very old specimens may grow to the height of 30 ft. or more. The other genera are smaller, and some have short stems and a very fern-like appearance, as, for example, the genus *Stangeria*, which was supposed to be a fern when it was first discovered and before fruiting specimens had been seen.

The large compound leaves are all borne directly on the main stem, generally in a single rosette at its apex, and as they die off they leave their fleshy leaf bases, which cover the stem and remain for an almost indefinite number of years.

Fig. 74.—Plant of *Cycas*, showing the main stem with the crown of leaves and the irregular branches which come on an old plant

The wood of the main trunks differs from that of the other Gymnosperms in being very loosely built, with a large pith and much soft tissue between the radiating bands of wood. There is a cambium which adds zones of secondary tissue, but it does not do its work regularly, and the cross section of an old Cycad stem shows disconnected rings of wood, accompanied by much soft tissue. The cells of the wood have bordered pits on their walls, and in the main axis the wood is usually all developed in a centrifugal direction, but in the axis of the cones some centripetal wood is found (refer to *c*, fig. 65, p. 97).

In their fructifications the Cycads stand even further apart from the rest of the Gymnosperms. One striking point is the enormous size of their *male* cones. The male cones consist of a stout axis, round which are spiral series of closely packed simple scales covered with pollen-bearing sacs (which bear no inconsiderable likeness to fern sporangia), the whole cone reaching 1½ ft. in length in some genera, and weighing several pounds. All the other Gymnosperms, except the Araucareæ, where they are an inch or two long, have male cones but a fraction of an inch in length.

In all the members of the family, excepting *Cycas* itself, the female fructifications also consist of similarly organized cones bearing a couple of seeds on each scale instead of the numerous pollen sacs. In *Cycas* the male cones are like those of the other genera, and reach an enormous size; but there are no female cones, for the seeds are borne on special leaflike scales. These are illustrated in fig. 75, which shows also that there are not two seeds (as in the other genera with cones) to each scale, but an indefinite number.

The leafy nature of the seed-bearing scale is an important and interesting feature. Although theoretically botanists are accustomed to accept the view that seeds are always borne on specially modified leaves (so that to a botanist even the "shell" of a pea-pod and the box of a poppy capsule are leaves), yet in *Cycas* alone among living plants are seeds really found growing on a large structure which has the appearance of a leaf. Hence, from this point of view (see p. 45, however, for a caution against concluding that the whole plant is similarly lowly organized), *Cycas* is the most primitive of all the living plants that bear seeds, and hence presumably the likest to the fossil ancestors of the seed-bearing types. In this character it is more primitive than the fossil group of

Fig. 75.— Seed-bearing Scale of *Cycas*, showing its lobed and leaflike character

s, Seeds attached on either side below the divisions of the sporophyll.

the *Cordaiteæ*, and comes very close to an intermediate group of fossils to be considered in the next chapter.

To enter into the detailed anatomy of the seeds would lead us too far into the realms of the specialist, but we must notice one or two points about them. Firstly, their very large size, for ripe seeds of *Cycas* are as large as peaches (and peaches, it is to be noted, are fruits, not seeds), and particularly the large size they attain *before* they are fertilized and have an embryo. Among the higher plants the young seeds remain very minute until an embryo is secured by the act of fertilization, but in the Cycads the seeds enlarge and lay in a big store of starch in the endosperm before the embryo appears, so that in the cases in which fertilization is prevented large, sterile "seeds" are nevertheless produced. This must be looked on as a want of precision in the mechanism, and as a wasteful arrangement which is undeniably primitive. An even more wasteful arrangement appears to have been common to the "seeds" of the Palæozoic period, for, though many fossil "seeds" are known in detail from the old rocks, not one is known to have any trace of an embryo. A general plan of the *Cycas* seed is shown in fig. 76, which should be compared with that of *Ginkgo* (fig. 68). The large size of the endosperm and the thick and complex seed-coats are characteristic features of both these structures. Another point that makes the Cycad seeds of special interest is the fact that the male cells (as in *Ginkgo*) are developed as active, free-swimming

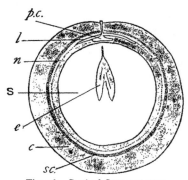

Fig. 76.—Seed of *Cycas* cut open

n, The nucellus, fused at the level *l* to the coat *c*; *sc*, stony layer of coat; *p.c*, pollen chamber in apex of the nucellus; s, "spore", filled with endosperm, in which lies the embryo *e*.

PAST HISTORIES OF PLANT FAMILIES

sperms, which swim towards the female cell in the space provided for them in the seed (see *p.c*, fig. 76).

The characters of the Cycads as they are now living prove them to be an extremely primitive group, and therefore presumably well represented among the fossils; and indeed among the Mesozoic rocks there is no lack of impressions which have been described as the leaves of Cycads. There is, however, very little reliable material, and practically none which shows good microscopic structure. Leaf impressions alone are most unsafe—more unsafe in this group, perhaps, than in any other—for reasons that will be apparent later on, and the conclusions that used to be drawn about the vast number of Cycads which inhabited the globe in the early Mesozoic must be looked on with caution, resulting from the experience of recent discoveries proving many of these leaves to belong to a different family.

There remain, however, many authentic specimens which show that *Cycas* certainly goes back very far in history, and specimens of this genus are known from the older Mesozoic rocks. We cannot say, however, as securely as used to be said, that the Mesozoic was the "Age of Cycads", although it was doubtless the age of plants which had much of the external appearance of Cycads.

From the Palæozoic we have no reliable evidence of the existence of Cycads, though the plants of that time included a group which has an undoubted connection with them.

Indeed, so far as fossil evidence goes, we must suppose that the Cycads, since their appearance, possibly at the close of the Palæozoic, have never been a dominant or very extensive family, though they grew in the past all over the world, and in Europe seem to have remained till the middle of the Tertiary epoch.

CHAPTER XII

PAST HISTORIES OF PLANT FAMILIES

V. Pteridosperms

This group consists entirely of plants which are extinct, and which were in the height of their development in the Coal Measure period. As a group they are the most recently discovered in the plant world, and but a few years ago the name "Pteridosperm" was unknown. They form, however, both one of the most interesting of plant families and one of the most numerous of those which flourished in the Carboniferous period.

To mention first the vital point of interest in their structure, they show *leaves which in all respects appear like ordinary foliage leaves, and yet bear seeds.* These leaves, which we now know bore the seeds, had long been considered as typical fern leaves, and had been named and described as fern leaves. There are two extremely important results from the discovery of this fossil group, viz. that leaves, to all appearance like ordinary foliage, can directly bear seeds, and that the leaves, though like fern leaves, bore seeds like those of a Cycad.

As the name *Pteridosperm* indicates, the group is a link between the ferns and the seed-bearing plants, and as such is of special interest and value to botanists.

The gradual recognition of this group from among the numerous plant fragments of Palæozoic age is one of the most interesting of the accumulative discoveries of fossil botany. Ever since fossil remains attracted the attention of enquiring minds many "ferns" have been recognized among the rich impressions of the Coal Measures. Most of them, however, were not connected with any structural material, and were given many different names of specific value. So numerous were these fern "species" that it was supposed that in the Coal

PAST HISTORIES OF PLANT FAMILIES

Measure period the ferns must have been the dominant class, and it is often spoken of even yet as the "Age of Ferns". From the rocks of the same age, preserved with their microscopical structure perfect, were stems which were called *Lyginodendron*. In the coal balls associated with these stems (which were the commonest of the stems so preserved) were also roots, petioles, and

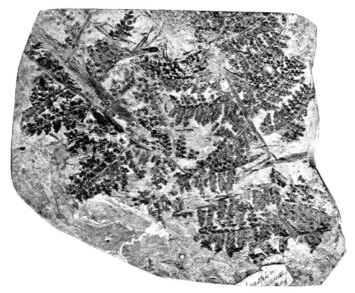

Fig. 77.—*Sphenopteris* Leaf Impression, the fernlike foliage of *Lyginodendron*

leaflets, but they were isolated, like the most of the fragments in a coal ball, and to each was given its name, with no thought of the various fragments having any connection with each other. Gradually, however, various fragments from the coal balls had been recognized as belonging together; one specimen of a petiole attached to a stem sufficed to prove that all the scattered petioles of the same type belonged also to that kind of stem, and when leaves were found attached to an isolated fragment of the petiole, the chain of proof was complete that the

ANCIENT PLANTS

leaves belonged to the stem, and so on. By a series of lengthy and painstaking investigations all the parts of the plant now called *Lyginodendron* have been brought together, and the impressions of its leaves have been connected with it, these being of the fernlike type so long called *Sphenopteris*, illustrated in fig. 77.

The anatomy of the main stem is very suggestive of that of a Cycad. The zones of secondary wood are

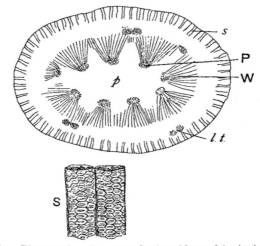

Fig. 78A.—Diagram of the Transverse Section of Stem of the *Lyginodendron*

p, Pith; P, primary wood groups; W, secondary wood; *l.t*, leaf trace; *s*, sclerized bands in the cortex; S, longitudinal view of wood elements to show the rows of bordered pits.

loosely built, the quantity of soft tissue between the radiating bands of wood, and the size of the pith being large, while from the main axis double strands of wood run out to the leaf base. The primary bundles, however, are not like those of a Cycad stem, but have groups of *centripetal* wood within the protoxylem, and thus resemble the primary bundles of *Poroxylon* (see p. 97), which are more primitive in this respect than those of the Cycads.

PAST HISTORIES OF PLANT FAMILIES

The roots of *Lyginodendron*, when young, were like those of the Marattiaceous ferns, their five-rayed mass of wood being characteristic of that family, and different from the type of root found in most other ferns (cf. fig. 78B with fig. 35 on p. 60). Unlike fern roots of any kind, however, they have well-developed zones of secondary

Fig. 78B.—Transverse Section of Root of *Lyginodendron*
w, Five-rayed mass of primary wood; *s*, zone of secondary wood; *c*, cortical and other soft tissues.

wood, in which they approach the Gymnospermic roots (see fig. 78B, *s*).

A further mixture of characters is seen in the vascular bundles of the petioles. A double strand, like that in the lower Gymnosperms, goes off to the leaf base from the main axis, but in the petiole itself the bundle is like a normal fern stele, and shows no characters in transverse section which would separate it from the ferns. Such a petiole is illustrated in fig. 79, with its **V**-shaped fernlike stele. On the petioles and stems were certain rough, spiny structures of the nature of complex hairs. In some cases they are glandular, as is seen in *g* in fig. 79, and as they seem to be unique in their appearance they have been of great service in the identification of the various isolated organs of the plant.

As is seen from fig. 77, the leaves were quite fern-like, but in structural specimens they have been found with the characteristic glandular hairs of the plant.

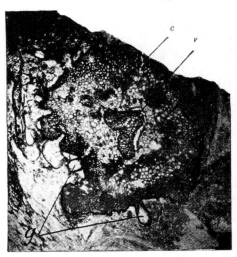

Fig. 79.—Transverse Section through Petiole of *Lyginodendron*

v, Fern-like stele; *c*, cortex; *g*, glandular hairlike protuberances.

The seeds were so long known under the name of *Lagenostoma* that they are still called by it, though they have been identified as belonging to *Lyginodendron*. They were small (about $\frac{1}{4}$ in. in maximum length) when compared with those of most other plants of the group, or of the Cycads, with which they show considerable affinity. They are too complex to describe fully, and have been mentioned already (see p. 76), so that they will not be described in much detail here. The diagrammatic figure (fig. 56) shows the essential characters of their longitudinal section, and their transverse section, as illustrated in fig. 80, shows the complex and elaborate mechanism of the apex.

Round the "seed" was a sheath, something like the husk round a hazel nut, which appears to have had the function of a protective organ, though what its real morphological nature may have been is as yet an unsolved problem. On the sheath were glandular hairs like those found on the petiole and leaves, which were, indeed, the first clues that led to the discovery of the connection between the seed and the plant *Lyginodendron*.

The pollen grains seem to have been produced in sacs very like fern sporangia developed on normal foliage leaves, each grain entered the cavity *pc* in the seed (see fig. 56), but of the nature of the male cell we are ignorant. In none of the fossils has any embryo been found in the "seeds", so that presumably they ripened, or at least matured their tissues, before fertilization.

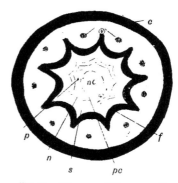

These, in a few words, are the essentials of the structures of *Lyginodendron*. But this plant is only one of a group, and at least two other of the Pteridosperms deserve notice, viz. *Medullosa*, which is more complex, and *Heterangium*, which is simpler than the central type.

Fig. 80.—Diagram of Transverse Section of *Lagenostoma* Seed near the Apex, showing the nine flutings *f* of the coat *c*; *v*, the vascular strand in each; *nc*, cone of nucellar tissue standing up in the fluted apex of the nucellus *n*; *pc*, the pollen chamber with a few pollen grains; *s*, space between nucellus and coat. Compare with diagram 56.

Heterangium is found also in rocks rather older than the coal series of England, though of Carboniferous age, viz. in the Calciferous sandstone series of Scotland, it occurs also in the ordinary Coal Measure nodules It is in several respects more primitive than *Lyginodendron*, and in particular in the structure of its stele comes nearer to that of ferns. The stele is, in fact, a solid mass of primary wood and wood parenchyma, corresponding in some degree to the protostele of a simple type (see p. 61, fig. 36), but it has towards the outside groups of protoxylem surrounded by wood in both centripetal and centrifugal directions, which are just like the primary bundles in *Lyginodendron*. Outside the primary mass of wood is a zone of secondary wood, but the quantity is not large in proportion to it (see fig. 81), as is common in Lyginodendron.

Though the primary mass is so fernlike in appearance the larger tracheids show series of bordered pits, as do most of the tracheids of the Pteridosperms, in which they show a Gymnosperm-like character.

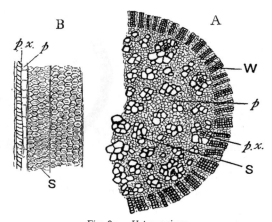

Fig. 81.—*Heterangium*

A, Half of the stele of a stem, showing the central mass of wood s mixed with parenchyma p. The protoxylem groups $p.x.$ lie towards the outside of the stele. Surrounding it is the narrow zone of small-celled secondary wood W. B, A few of the wood cells in longitudinal view: $p.x.$, Protoxylem; p, parenchyma. s, Large vessels with rows of bordered pits.

The foliage of *Heterangium* was fernlike, with much-divided leaves similar to those of *Lyginodendron*. We have reason to suspect, though actual proof is wanting as yet, that small Gymnosperm-like seeds were borne directly on these leaves.

Medullosa has been mentioned already (see p. 72) because of the interesting and unusually complex type of its vascular anatomy. Each individual stele of the group of three in the stem, however, is essentially similar to the stele of a Heterangium.

PAST HISTORIES OF PLANT FAMILIES

Though the whole arrangement appears to differ so widely from other stems in the plant world, careful comparison with young stages of recent Cycads has indicated a possible remote connection with that group, while in the primary arrangements of the protosteles a likeness may be traced to the ferns. The roots, even in their primary tissues, were like those of Gymnosperms, but the foliage with its compound leaves was quite fern-like externally. A small part of a leaf is shown in fig. 83, and is clearly like a fern in superficial appearance. The leaves were large, and the leaf bases strong and well supplied with very numerous branching vascular bundles.

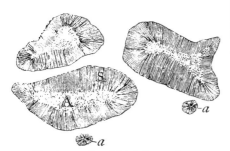

Fig. 82.—Steles of *Medullosa* in Cross Section of the Stem

A, Primary solid wood; S, surrounding secondary wood.

Fig. 83.—Part of a Leaf of *Medullosa*, known as *Alethopteris*, for long supposed to be a Fern

The connection between this plant and certain large three-ribbed seeds known as *Trigonocarpus* is strongly suspected, though actual continuity is not yet established in any of the specimens hitherto discovered. These seeds have been mentioned before (p. 76 and p. 82). They were larger than the other fossil seeds which we have mentioned, and, with their fleshy coat, were similar in general organization to the Cycads, though the fact that the seed coat stood free from the inner tissues right down to the base seems to mark them as being more primitive (cf. fig. 55, p. 76).

Of impressions of the Pteridosperms the most striking is, perhaps, the foliage known as *Neuropteris* (see fig. 6, p. 13), to which the large seeds are found actually attached (cf. fig. 85).

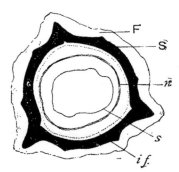

Fig. 84.—Diagrammatic Section of a Transverse Section of a Seed of *Trigonocarpus*

s, Stone of coat with three main ridges and six minor ones. F, Flesh of coat: *if*, inner flesh ; *n*, nucellus, crushed and free from coat; *s*, spore wall.

Fig. 85.—Fragment of Foliage of *Neuropteris* with Seed attached, showing the manner in which the seeds grew on the normal foliage leaves in the Pteridosperms

Ever-increasing numbers of the "ferns" are being recognized as belonging to the Pteridosperms, but *Heterangium*, *Lyginodendron*, and *Medullosa* form the three principal genera, and are in themselves a series indicating the connection between the fernlike and Cycadean characters.

Before the fructifications were suspected of being seeds the anatomy of these plants was known, and

PAST HISTORIES OF PLANT FAMILIES

their nature was partly recognized from it alone, though at that time they were supposed to have only fernlike spores.

The very numerous impressions of their fernlike foliage from the Palæozoic rocks indicate that the plants which bore such leaves must have existed at that time in great quantity. They must have been, in fact, one of the dominant types of the vegetation of the period. The recent discovery that so large a proportion of them were not ferns, but were seed-bearing plants, alters the long-established belief that the ferns reached their high-water mark of prosperity in the Coal Measure period. Indeed, the fossils of this age which remain undoubtedly true ferns are far from numerous. It is the seed-bearing Pteridosperms which had their day in Palæozoic times. Whether they led directly on to the Cycads is as yet uncertain, the probability being rather that they and the Cycads sprang from a common stock which had in some measure the tendencies of both groups.

That the Pteridosperms in themselves combined many of the most important features of both Ferns and Gymnosperms is illustrated in the account of them given above, which may be summarized as follows:—

SALIENT CHARACTERS OF THE PTERIDOSPERMS

Gymnospermic	*Fernlike*
	Primary structure of root.
Secondary thickening of root.	
	In *Heterangium* and *Medullosa* the solid centripetal primary wood of stele.
Pits on tracheæ of primary wood. Secondary thickening of stem. Double leaf trace.	
	Fernlike stele in petiole. Fernlike leaves. Sporangia pollen-sac-like. Reproductive organs borne directly on ordinary foliage leaves.
General organization of the seed.	

Thus it can be seen at a glance, without entering into minutiæ, that the characters are divided between the two groups with approximate equality. The connection with Ferns is clear, and the connection with Gymnosperms is clear. The point which is not yet determined, and about which discussion will probably long rage, is the position of this group in the whole scheme of the plant world. Do they stand as a connecting link between the ferns on one hand and the whole train of higher plants on the other, or do they lead so far as the Cycads and there stop?

CHAPTER XIII

PAST HISTORIES OF PLANT FAMILIES

VI. The Ferns

Unfortunately the records in the rocks do not go back so far as to touch what must have been the most interesting period in the history of the ferns, namely, the point where they diverged from some simple ancestral type, or at least were sufficiently primitive to give indications of their origin from some lower group.

Before the Devonian period all plant impressions are of little value, and by that early pre-Carboniferous time there are preserved complex leaves, which are to all appearance highly organized ferns.

To-day the dominant family in this group is the *Polypodiaceæ*. It includes nearly all our British ferns, and the majority of species for the whole world. This family does not appear to be very old, however, and it cannot be recognized with certainty beyond Mesozoic times.

From the later Mesozoic we have only material in the form of impressions, from which it is impossible to

PAST HISTORIES OF PLANT FAMILIES 125

draw accurate conclusions unless the specimens have sporangia attached to them, and this is not often the case. The cuticle of the epidermis or the spores can sometimes be studied under the microscope after special treatment, but on the whole we have very little information about the later Mesozoic ferns.

A couple of specimens from the older Mesozoic have been recently described, with well-preserved structure, and they belong to the family of the Osmundas (the so-called "flowering ferns", because of the appearance of special leaves on which all the sporangia are crowded), and show in the anatomical characters of their stems indications that they may be related to an old group, the *Botryopterideæ*, in which are the most important of the Palæozoic ferns.

In the Palæozoic rocks there are numerous impressions as well as fern petrifactions, but in the majority of cases the connection between the two is not yet established. There were two main series of ferns, which may be classed as belonging to

 I. Marattiaceæ.
 II. Botryopterideæ.

Of these the former has still living representatives, though the group is small and unimportant compared with what it once was; the latter is entirely extinct, and is chiefly developed in the Carboniferous and succeeding Permian periods.

The latter group is also the more interesting, for its members show great variety, and series may be made of them which seem to indicate the course taken in the advance towards the Pteridosperm type. For this reason the group will be considered first, while the structure of the Pteridosperms is still fresh in our minds.

The Botryopterideæ formed an extensive and elaborate family, with its numerous members of different degrees of complexity. There is, unfortunately, but little known as to their external appearance, and almost no definite information about their foliage. They are principally known by the anatomy of their stems and petioles. Some of them had upright trunks like small tree ferns (living tree ferns belong to quite a different family, however), others appear to have had underground stems, and many were slender climbers.

Fig. 86.—Stele of *Asterochlaena*, showing its deeply lobed nature

In their anatomy all the members of the family have monostelic structure (see p. 62). This is noteworthy, for at the present time though a number of genera are monostelic, no family whose members reach any considerable size or steady growth is exclusively monostelic. In the shape of the single stele, there is much variety in the different genera, some having it so deeply lobed that only a careful examination enables one to recognize its essentially monostelic nature. In fig. 86 a radiating star-shaped type is illustrated. Between this elaborate type of protostele in *Asterochlaena*, and the simple solid circular mass seen in *Botryopteris* itself (fig. 88) are all possible gradations of structure.

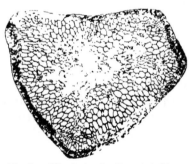

Fig. 87.—The Stele of a Botryopteridean Stem, showing soft tissue in the centre of the solid wood of the protostele. (Microphoto.)

In several of the genera the centre of the wood is not entirely solid, but has cells of soft tissue, an incipient pith, mixed with scattered tracheids, as in fig. 87.

In most of the genera numerous petioles are given off from the main axis, and these are often of a large size compared with it, and may sometimes be thicker

PAST HISTORIES OF PLANT FAMILIES

than the axis itself. Together with the petiole usually come off adventitious roots, as is seen in fig. 88, which shows the main axis of a *Botryopteris*. The petioles of the group show much variety in their structure, and some are extremely complex. A few of the shapes assumed by the steles of the petioles are seen in fig. 89; they are not divided into separate bundles in any of the known forms, as are many of the petiole steles of other families.

In one genus of the family secondary wood has been observed.

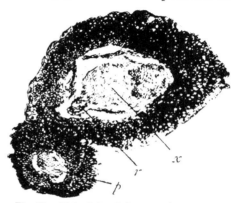

Fig. 88.—Main Axis of *Botryopteris* with simple solid Protostele x. A petiole about to detach itself p and the strand going out to an adventitious root r are also seen. (Micro-photograph.)

This is highly suggestive of the condition of the stele in *Heterangium*, where the large mass of the primary wood is surrounded by a relatively small quan-

Fig. 89.—Diagrams showing the Shapes of the Steles in some of the Petioles of different Genera of Botryopterideæ

A, *Zygopteris*; B, *Botryopteris*; C, *Tubicaulis*; D, *Asterochlaena*.

tity of secondary thickening, developed in normal radial rows from a cambium.

Another noteworthy point in the wood of these plants is the thickening of the walls of the wood cells. Many of them have several rows of bordered pits, and are, individually, practically indistinguishable from those of the Pteridosperms, cf. fig. 81 and fig. 90. These are unlike the characteristic wood cells of modern ferns and of the other family of Palæozoic ferns.

The foliage of most members of the family is unknown, or at least, of the many impressions which possibly belong to the different genera, the most part have not yet been connected with their corresponding structural material. There are indications, however, that the leaves were large and complexly divided.

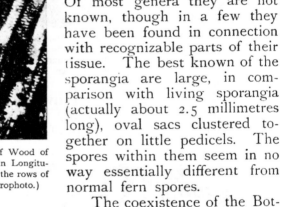

Fig. 90.—Tracheæ of Wood of Botryopteridean Fern in Longitudinal Section, showing the rows of pits on the walls. (Microphoto.)

The fructifications were presumably fern sporangia of normal but rather massive type. Of most genera they are not known, though in a few they have been found in connection with recognizable parts of their tissue. The best known of the sporangia are large, in comparison with living sporangia (actually about 2.5 millimetres long), oval sacs clustered together on little pedicels. The spores within them seem in no way essentially different from normal fern spores.

The coexistence of the Botryopterideæ and Pteridosperms, and the several points in the structure of the former which seem to lead up to the characters of the latter group, are significant. The Botryopterideæ, even were they an entirely isolated group, would be interesting from the variety of structures and the variations of the monostele in their anatomy; and the prominent place they held in the Palæozoic flora, as the greatest family of ferns of that period, gives them an important position in fossil botany.

The other family of importance in Palæozoic times,

PAST HISTORIES OF PLANT FAMILIES 129

the MARATTIACEÆ, has descendants living at the present day, though the family is now represented by a small number of species belonging to but five genera which are confined to the tropics. Perhaps the best known of these is the giant "Elephant Fern", which sends up from its underground stock huge complex fronds ten or a dozen feet high. Other species are of the more usual size and appearance of ferns, while some have sturdy trunks above-ground supporting a crown of leaves. The members of this family have a very complex anatomy, with several series of steles of large size and irregular shape. Their fructifications are characteristic, the sporangia being placed in groups of about five to a dozen, and fused together instead of ripening as separate sacs as in the other fern families.

Impressions of leaves with this type of sorus (group of fern sporangia) are found in the Mesozoic rocks, and these bridge over the interval between the living members of the family and those which lived in Palæozoic times.

In the Coal Measure and Permian periods these plants flourished greatly, and there are remains of very numerous species from that time. The family was much more extensive then than it is now, and the individual members also seem to have reached much greater dimensions, for many of them had the habit of large tree ferns with massive trunks. Up till Triassic times half of the ferns appear to have belonged to this family; since then, however, they seem to have dwindled gradually down to the few genera now existing.

On the Continent fossils of this type with well-preserved structure have long been known to the general public, as their anatomy gave the stones a very beautiful appearance when polished, so that they were used for decorative purposes by lapidaries before their scientific interest was recognized.

The members of the Palæozoic Marattiaceæ which have structure preserved generally go by the generic name *Psaronius*, in which there is a great number of species. They show considerable uniformity in their essential structure (in which they differ noticeably from the group of ferns just described), so that but one type will be considered.

In external appearance they probably resembled the "tree ferns" of the present day (though these belong to an entirely different family), with massive stumps, some of which reached a height of 60 ft. The large spreading leaves were arranged in various ways on the stem, some in a double row along it, as is seen by the impressions of the leaf scars, and others in complex spirals. On the leaves were the spore sacs, which were in groups, some completely fused like those of the modern members of the family, and others with independent sporangia massed in well-defined groups. In their microscopic structure also they appear to have been closely similar to those of the living Marattiaceæ.

The transverse section of a stem shows the most characteristic and best-known view of the plant. This is shown in fig. 91, in somewhat diagrammatic form.

The mass of rootlets which entirely permeate and surround the outer tissues of the stem is a very striking and characteristic feature of all the species of *Psaronius*. Though such a mass of roots is not found in the living species, yet the microscopic structure of an individual fossil root is almost identical with that of a living *Marattia*.

Though these plants were so successful and so important in Palæozoic times, the group even then seems to have possessed little variety and little potentiality for advance in new directions. They stand apart from the other fossils, and the few forms which now compose the living Marattiaceæ are isolated from the present successful types of modern ferns. From the *Psaronieæ* we can trace no development towards a modern series of plants, no connection with another important group in the past. They appear to have culminated in the later Palæozoic and to have slowly dwindled ever since. It has been suggested that the male fructifications of the Bennettiteæ and the Pteridosperms show some likeness to the Marattiaceæ, but there does not seem much to support any view of phylogenetic connection between them.

Before leaving the palæozoic ferns, mention should be made of the very numerous leaf impressions which

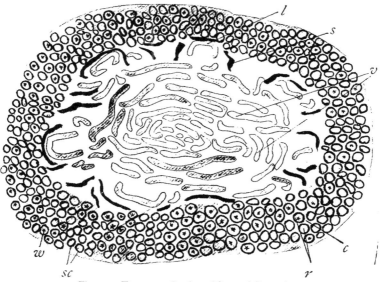

Fig. 91.—Transverse Section of Stem of *Psaronius*

v, Numerous irregularly-shaped steles; *s*, irregular patches of sclerenchyma; *l*, leaf trace going out as a horseshoe-shaped stele; *c*, zone of cortex with numerous adventitious roots *r* running through it; *sc*, sclerized cortical zone of roots; *w*, vascular strand of roots.

seem to show true fern characters, though they have not been connected with material showing their internal structure. Among them it is rare to get impressions with the *sori* or sporangia, but such are known and are in themselves enough to prove the contention that true ferns existed in the Palæozoic epoch. For it might be mentioned as a scientific curiosity, that after the discovery that so many of the leaf impressions which had always been supposed to be ferns, really belonged to the seed-bearing Pteridosperms, there was a period of panic among some botanists, who brought forward the startling idea that there were *no* ferns at all in the Palæozoic periods, and that modern ferns were degenerated seed-bearing plants!

These two big groups from the Palæozoic include practically all the ferns that then flourished. They have been spoken of (together with a few other types of which little is known) as the *Primofilices*, a name which emphasizes their primitive characters. As can be seen by the complex organization of the genera, however, they themselves had advanced far beyond their really primitive ancestors. There is clear indication that the Botryopterideæ were in a period of change, what might almost be termed a condition of flux, and that from their central types various families separated and specialized. Behind the Botryopterideæ, however, we have no specimens to show us the connection between them and the simpler groups from which they must have sprung. From a detailed comparative study of plant anatomy we can deduce some of the essential characters of such ancestral plants, but here the realm of fossil botany ceases, to give place to theoretical speculation. As a fact, there is a deep abyss between the ferns and the other families of the Pterido-

Fig. 92.—Impression of Palæozoic Fern, showing *sori* on the pinnules. (Photo.)

PAST HISTORIES OF PLANT FAMILIES

phytes, which is not yet bridged firmly enough for any but specialists, used to the hazardous footing on such structures, to attempt to cross it. Until the buttresses and pillars of the bridge are built of the strong stone of fossil structures we must beware of setting out on what would prove a perilous journey.

In the Coal Measures and previous periods we see the ferns already represented by two large families, differing greatly from each other, and from the main families of modern ferns which sprung at a later date from some stock which we have not yet recognized. But though their past is so obscure, the palæozoic ferns and their allies throw a brilliant light on the course of evolution of the higher groups of plants, and the gulf between ferns and seed-bearing types may be said to be securely bridged by the Botryopterideæ and the Pteridosperms.

CHAPTER XIV

PAST HISTORIES OF PLANT FAMILIES

VII. The Lycopods

The present-day members of this family are not at all impressive, and in their lowliness may well be overlooked by one who is not interested in unpretending plants. The fresh green mosslike *Selaginella* grown by florists as ornamental borders in greenhouses and the creeping "club moss" twining among the heather on a Highland moor are probably the best known of the living representatives of the Lycopods. In the past the group held a very different position, and in the distant era of the Coal Measures it held a dominant one. Many of the giants of the forest belonged to the family (see frontispiece), and the number of species it contained was very great.

Let us turn at once to this halcyon period of the group. The history of the times intervening between it and the present is but the tale of the dying out of the large species, and the gradual shrinking of the family and dwarfing of its representative genera.

It is difficult to give the characters of a scientific family in a few simple words; but perhaps we may describe the living Lycopods as plants with creeping stems which divide and subdivide into two with great regularity, and which bear large numbers of very small pointed leaves closely arranged round the stem. The fruiting organs come at the tips of the branches, and sometimes themselves divide into two, and in these cone-like axes the spore cases are arranged, a single one on the upper side of each of the scales (see p. 67, fig. 46, A). In the Lycopods the spores are all alike, in the Selaginellas there are larger spores borne in a small number (four) in some sporangia (see fig. 53, p. 75), and others in large numbers and of smaller size on the scales above them. The stems are all very slender, and have no zones of secondary wood. They generally creep or climb, and from them are put out long structures something like roots in appearance, which are specially modified stem-like organs giving rise to roots.

From the fossils of the Coal Measures *Lepidodendron* must be chosen as the example for comparison. The different species of this genus are very numerous, and the various fossilized remains of it are among the commonest and best known of palæontological specimens. The huge stems are objects of public interest, and have been preserved in the Victoria Park in Glasgow in their original position in the rocks, apparently as they grew with their spreading rootlike organs running horizontally. A great stump is also preserved in the Manchester Museum, and is figured in the frontispiece. While among the casts and impressions the leaf bases of the plant are among the best preserved and the most beautiful (see fig. 93). The cone has already been illus-

PAST HISTORIES OF PLANT FAMILIES 135

trated (see fig. 46 and fig. 9), and is one of the best known of fossil fructifications.

From the abundant, though scattered material, fossil botanists have reconstructed the plants in all their detail. The trunks were lofty and of great thickness, bearing

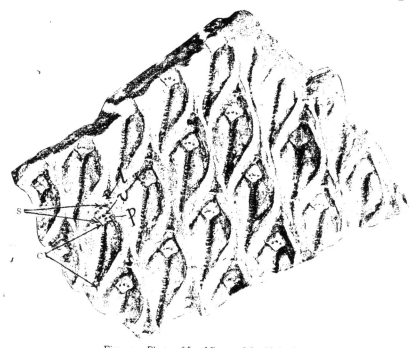

Fig. 93.—Photo of Leaf Bases of *Lepidodendron*

C, Scar of leaf; S, leaf base. In the scar: *v*, mark of severed vascular bundle, and *p*, of parichnos. *l*, Ligule scar.

towards the apex a much-branched crown, the branches, even down to the finest twigs, all dividing into two equal parts. The leaves, as would be expected from the great size of the plants, were much bigger than those of the recent species (fig. 93 shows the actual size of the leaf bases), but they were of the same *relatively* small size as compared with the stems, and of the same simple pointed shape. A transverse section across the apex of a fertile branch shows these closely packed leaves arranged in series round the axis, those towards the outside show the central vascular strand which runs through each.

The markings left on the well-preserved leaf-scars

Fig. 94.—Section across an Axis surrounded by many Leaves, which shows their simple shape and single central vascular bundle v

indicate the main features of the internal anatomy of the leaves. They had a single central vascular strand (v, fig. 93), on either side of which ran a strand of soft tissue p called the parichnos, which is characteristic of the plants of this group. While another similarly obscure structure associated with the leaf is the little scale-like ligule l on its upper surface.

The anatomy of the stems is interesting, for in the

PAST HISTORIES OF PLANT FAMILIES 137

different species different stages of advance are to be found, from the simple solid protostele with a uniform mass of wood to hollow ring steles with a pith. An interesting intermediate stage between these two is found in *Lepidodendron selaginoides* (see fig. 95), where the central cells of the wood are not true water-conducting

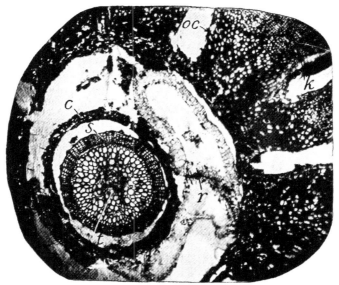

Fig. 95.—Transverse Section of *Lepidodendron selaginoides*, showing the circular mass of primary wood, the central cells of which are irregular water-storage tracheides

s, Zone of secondary wood; *c*, inner cortical tissues; *r*, intrusive burrowing rootlet; *oc*, outer cortical tissues with corky external layers *k*. (Microphoto.)

cells, but short irregular water-storage tracheides (see p. 56), which are mixed with parenchyma. All the genera of these fossils have a single central stele, round which it is usual to find a zone of secondary wood of greater or less extent according to the age of the plant.

Some stems instead of this compact central stele have a ring of wood with an extensive pith. Such a type is illustrated in fig. 96, which shows but a part of the circle of wood, and the zone of the secondary wood outside it, which greatly exceeds the primary mass in thickness. This zone of secondary wood became very extensive in old stems, for, as will be imagined, the primary wood was not sufficient to supply the large trunks. The method of its development from a normal cambium in radiating rows of uniform tracheides is quite similar to that which is found in the pines to-day. This is the

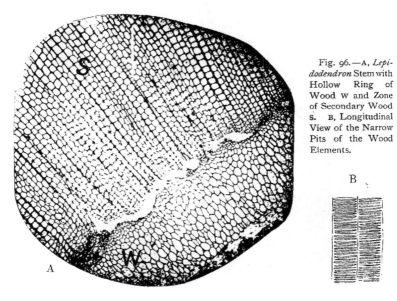

Fig. 96.—A, *Lepidodendron* Stem with Hollow Ring of Wood w and Zone of Secondary Wood s. B, Longitudinal View of the Narrow Pits of the Wood Elements.

most important difference between the living and the fossil stems of the family, for no living plants of the family have such secondary wood. On the other hand, the individual elements of this wood are different from those of the higher families hitherto considered, and have narrow slit-like pits separated by bands of thickening on the longitudinal walls. Such tracheides are found commonly in the Pteridophytes, both living and fossil. Their type is seen in fig. 96, B, which should be compared with that in figs. 78, A and 62, B to see the contrast with the higher groups.

PAST HISTORIES OF PLANT FAMILIES 139

To supply the vascular tissues of the leaf traces, simple strands come off from the outer part of the primary wood, where groups of small-celled protoxylem project (see *px* in fig. 97). The leaf strands *lt* move out through the cortex in considerable numbers to supply the many leaves, into each of which a single one enters.

Fig. 97.—Transverse Section of Outer Part of Primary Wood of *Lepidodendron*, showing *px*, projecting protoxylem groups; *lt*, leaf trace coming from the stele and passing (as *lt*¹) through the cortex

As regards the fructifications of *Lepidodendron* much could be said were there space. The many genera of *Lepidodendron* bore several distinct types of cones of different degrees of complexity. In several of the genera the cones were simple in organization, directly comparable with those of the living Lycopods, though on a much larger scale (see p. 67). In some the spores were uniform, all developing equally in numerous tetrads. The sporophyll was radially extended, and along it the large sausage-shaped sporangia were attached (see fig. 98). The tips of the sporophylls overlapped and afforded protection to the sporangia. The axis of the cone had a central stele with wood elements like those in the stem. The appearance of a transverse section of an actual cone is shown in fig. 99. Here the sporangia are irregular in shape, owing to their contraction after ripeness and during fossilization. Other cones had sporangia similar in size and shape, but which produced spores of two kinds, large ones resulting from the ripening of only two or three tetrads in the lower sporangia, and numerous small ones in the sporangia above.

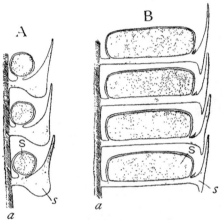

Fig. 98.—Longitudinal Diagram, showing the arrangement of the elongated sporangia on the sporophylls

a, Main axis, round which the sporophylls are inserted; s, sporangium; *s*, leaflike end of sporophyll.

The similarity between the *Lepidodendron* and the modern Lycopod cone has been pointed out already (p. 67), and it is this which forms the principal guarantee that they belong to the same family, though the size and wood development of the palæozoic and the modern plants differ so greatly.

The large group of the Lepidodendra included some members whose fructifications had advanced so far beyond the simple sporangial cones described above as to approach very closely to seeds in their construction. This type was described on p. 75, fig. 54, in a series of female fructifications, so that its essential structure need not be recapitulated.

PAST HISTORIES OF PLANT FAMILIES

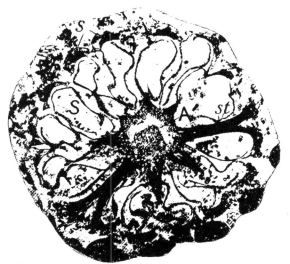

Fig. 99.—Transverse Section through Cone of *Lepidodendron*

A, Main axis with woody tissue; *st*, stalks of sporophylls cut in oblique longitudinal direction; *s*, tips of sporophylls cut across; s, sporangia with a few groups of spores. (Microphoto.)

The section shown in fig. 100 is that cut at right angles to that in which the sporangia are shown in fig. 98, viz. tangential to the axis. A remarkable feature of the plant is that there were also round those sporangia which bore the numerous small spores (corresponding to pollen grains) enclosing integument-like flaps similar to those shown in fig. 100, *sp. f.*

This type of fructification is the nearest approach to seed and pollen grains reached by any of the Pteridophytes, and its appearance at a time when the Lycopods were one of the dominant families

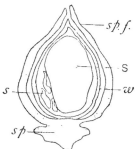

Fig. 100.—Section through one Sporangium of *Lepidocarpon*

sp, Sporophyll; *sp.f.*, flaps of sporophyll protecting sporangium; s, large spore within the sporangium wall *w*; *s*, the three aborted spores of the tetrad to which s belongs.

is suggestive of the effect that such a position has on the families occupying it, however lowly they may be. The simple Pteridophyte Lycopods had not only the tall trunks and solid woody structure of a modern tree, but also a semblance of its seeds. Whether this line of development ever led on to any of the higher families is still uncertain. The feeling of most specialists is that it did not; but there are not wanting men who support the view that the lycopod affinity evolved in time and entered the ranks of the higher plants, and indeed there are many points of superficial likeness between the palæozoic Lycopods and the Coniferæ. Judged from their internal structure, however, the series through the ferns and Pteridosperms leads much more convincingly to the seed plants.

In their roots; or rather in the underground structures commonly called roots, the Lepidodendrons were also remarkable. Even more symmetrically than in their above-ground branching, the base of their trunks divided; there were four main large divisions, each of which branched into two and these into two again. These structures were called *Stigmaria*, and were common to all species of *Lepidodendron* and also the group of *Sigillaria* (see fig. 102). On these horizontally running structures (well shown in the frontispiece) small appendages were borne all over their surface in great profusion, which were, both in their function and microscopic structure, rootlets. They left circular scars of a characteristic appearance on the big trunks, of which they were the only appendages. These scars show clearly on the fragments along the ledge to the left of the photograph. The exact morphological nature of the big axes is not known; their anatomy is not like that of roots, but is that of a stem, yet they do not bear what practically every stem, whether underground or not, has developed, namely leaves, or scales representing reduced leaves. Their nature has been commented on previously (p. 69), and we cannot discuss the point further, but must be

PAST HISTORIES OF PLANT FAMILIES 143

content to consider them as a form of root-bearing stem, practically confined to the Lycopods and principally developed among the palæozoic fossils of that group.

In microscopic structure the rootlets are extremely well known, because in their growth they have penetrated the masses of the tissues of other plants which were being petrified and have become petrified with them. The mass of decaying vegetable tissue on which the living plants of the period flourished were everywhere pierced by these intrusive rootlets, and they are found petrified inside otherwise perfect seeds, in the hearts of woody stems, in leaves and sporangia, and sometimes even inside each other! Fig. 95 shows such a root *r* lying in the space left by the decay of the soft tissue of the inner cortex in an otherwise excellently preserved *Lepidodendron* stem (see also fig. 101). In fig. 101 their simple structure is seen. They are often extremely irregular in shape, owing to the way they seem to have twisted and flattened themselves in order to fit into the tissues they were penetrating. No root hairs seem to have been developed in these rootlets, but otherwise their structure is that of a typical simple root, and very like the swamp-penetrating rootlets of the living Isoetes.

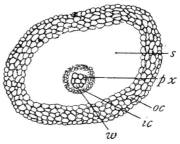

Fig. 101.—Transverse Section through a Rootlet of *Stigmaria*

oc, Outer cortex; *s*, space; *ic*, inner cortex; *w*, wood of vascular strand (wood only preserved); *px*, protoxylem group.

The Stigmarian axes and their rootlets are very commonly found in the "underclays" and "gannister" beds which lie below the coal seams (see p. 25), and they may sometimes be seen attached to a bit of the trunk growing upwards through the layers. They and the aerial stems of Lepidodendron are perhaps the commonest and most widely known of fossil plants.

144   ANCIENT PLANTS

Before leaving the palæozoic Lycopods another genus must be mentioned, which is also a widely spread and important one, though it is less well known than its contemporary. The genus *Sigillaria* is best known by its impressions and casts of stems covered by leaf scars. The stems were sometimes deeply ribbed, and the leaf scars were arranged in rows and were more or less hexa-

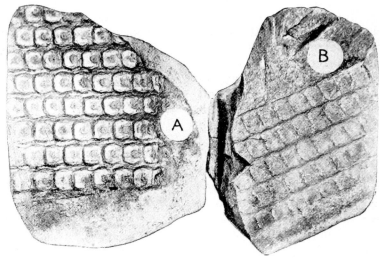

Fig. 102.—Cast and Reverse of Leaf Scars of *Sigillaria*. In A the shape of the leaf bases is clearly shown, the central markings in each being the scar of the vascular bundle and parichnos

gonal in outline, as is seen in fig. 102, which shows a cast and its reverse of the stem of a typical *Sigillaria*.

In its primary wood *Sigillaria* differed from *Lepidodendron* in being more remote from the type with a primary solid stele. Its woody structure was that of a ring, in some cases irregularly broken up into crescent-shaped bundles. The secondary wood was quite similar to that of *Lepidodendron*.

Stigmaria and its rootlets belong equally to the two plants, and hitherto it has been impossible to tell whether any given specimen of *Stigmaria* had belonged to a *Lepidodendron* or a *Sigillaria*. Between the two genera there certainly existed the closest affinity and similarity in general appearance.

These two genera represent the climax of development of the Lycopod family. In the Lower Mesozoic some large forms are still found, but all through the Mesozoic periods the group dwindled, and in the Tertiary little is known of it, and it seems to have taken the retiring position it occupies to-day.

CHAPTER XV

PAST HISTORIES OF PLANT FAMILIES

VIII. The Horsetails

The horsetails of to-day all belong to the one genus, *Equisetum*, among the different species of which there is a remarkably close similarity. Most of the species love swampy land, and even grow standing up through water; but some live on the dry clay of ploughed fields. Wherever they grow they usually congregate in large numbers, and form little groves together. They are easily recognized by their delicate stems, branching in bottle-brush fashion, and the small leaves arranged round them in whorls, with their narrow teeth joined to a ring at the base. At the end of some of the branches come the cones, with compactly arranged and simple sporophylls all of one kind. In England most plants of this family are but a few inches or a foot in height, though one species sometimes reaches 6 ft., while in South America there are groves of delicate-stemmed plants 20 ft. high.

The ribbed stems and the whorls of small, finely toothed leaves are the most important external characteristics of the plants, while in their internal anatomy the hollow stems have very little wood, which is arranged in a series of small bundles, each associated with a hollow canal in the ground tissue.

The family stands apart from all others, and even between it and the group of Lycopods there seems to be a big gap across which stretch no bonds of affinity. Has the group always been in a similar position, and stood isolated in a backwater of the stream of plant life?

Fig. 103.—Impression of Leaf Whorl of *Equisetites* from the Mesozoic Rocks, showing the narrow toothed form of the leaves. (Photo.)

In the late Tertiary period they seem to have held much the same position as they do now, and we learn nothing new of them from rocks of that age. When, however, we come to the Mesozoic, the members of the family are of greater size, though they appear (to judge from their external appearance) to have been practically identical with those now living in all their arrangements. In some beds their impressions are very numerous, but unfortunately most are without any indication of internal structure. Fossils from the Mesozoic are called *Equisetites*, a name which indicates that they come very close to the living ones in their characters. In the Lower Mesozoic some of these stems seem to have reached the great size of a couple of feet in circumference, but to have no essential difference from the others of the group.

When, however, we come to the Palæozoic rocks we find many specimens with their structure preserved, and we are at once in a very different position as regards the family.

First in the Permian we meet with the important

PAST HISTORIES OF PLANT FAMILIES 147

genus of plant called *Calamites*, which were very abundant in the Coal Measures. Many of the Calamites were of great size, for specimens with large trunks have been found 30 ft. and more long, which when growing must certainly have been much taller than that. The number of individuals must also have been very great, for casts and impressions of the genus are among the commonest

Fig. 104.—Small Branches attached to stouter Axis of *Calamites*.   Photo of Impression

fossils. They were, in fact, one of the dominant groups of the period. Like the Lycopods, the Equisetaceæ reached their high-water mark of development in the Carboniferous period; at that time the plants were most numerous, and of the largest size and most complicated structure that they ever attained.

As will be immediately suspected from analogy with the Lycopods, they differed from the modern members of the family in their strongly developed anatomy, and in the strength and quantity of their secondary wood.

Yet in their external appearance they probably resembled the living genus in all essentials, and the groves of the larger ones of to-day growing in the marshes probably have the appearance that the palæozoic plants would have had if looked at through a reversed opera glass.

Fig. 104 is a photograph of some of the small branches of a Calamite, in which the ribbed stem can be seen, and on the small side twigs the fine, pointed leaves lying in whorls.

In most of the fossil specimens, however, particularly the larger ones, the ribs are not those of the true surface, but are those marked on the *internal cast* of the pith.

Among tissue petrifactions there are many Calamite stems of various stages of growth. In the very young ones there are only primary bundles, and

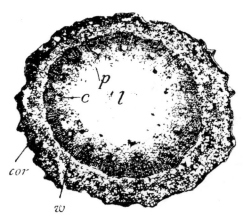

Fig. 105.—Transverse Section of *Calamites* Stem with Secondary Wood *w* formed in Regular Radial Rows in a Solid Ring

c, Canals associated with the primary bundles; *p*, cells of the pith, which is hollow with a cavity *l*. *cor*, Cortex and outer tissues well preserved. (Microphoto.)

these little stems are like those of a living Equisetum in their anatomy, and have a hollow pith and small vascular bundles with canals associated. The fossil forms, however, soon began to grow secondary wood, which developed in regular radial rows from a cambium behind the primary bundles and joined to a complete ring.

A stem in this stage of development is seen in fig. 105, where only the wood and internal tissues are preserved. The very characteristic canals associated with the primary bundles are clearly shown. The amount of secondary

PAST HISTORIES OF PLANT FAMILIES 149

wood steadily increased as the stems grew (there appear to have been no "annual rings") till there was a very large quantity of secondary tissue of regular texture, through which ran small medullary rays, so that the stems became increasingly like those of the higher plants as they grew older. It is the primary structure which is the important factor in considering their affinity, and that is essentially the same as in the other members of the family in which secondary thickening is not developed. As we have seen already in other groups of fossils, secondary wood appears to develop on similar lines whenever it is needed in any group, and therefore has but little value as an indication of systematic position.

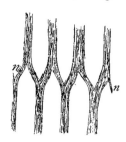

Fig. 106.—Diagram of the Arrangement of the Bundles at the Node of a *Calamite*, showing how those of consecutive internodes alternate

n, Region of node

This important fact is one, however, which has only been realized as a result of the study of fossil plants.

The longitudinal section of the stems, when cut tangentially, is very characteristic, as the bundles run straight down to each node and there divide, the neighbouring halves joining so that the bundles of each node alternate with those of the ones above and below it (see fig. 106).

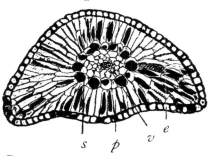

Fig. 107.—Leaf of *Calamites* in Cross Section

v, Vascular bundles; *s*, cells of sheath, filled with blackened contents; *p*, palisade cells; *e*, epidermis.

The leaves which were attached at the nodes were naturally much larger than those of the present Equisetums, though they were similarly simple and undivided. Their anatomy is preserved in a number of cases (see fig. 107), and was simple, with a single small strand of vascular tissue lying in the centre. They had certain large cells, sometimes very black in the fossils, which may have been filled with mucilage.

The young roots of these plants have a very characteristic cortex, which consists of cells loosely built together in a lacelike fashion, with large air spaces, so that they are much like water plants in their appearance

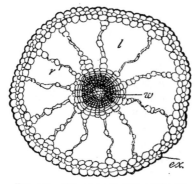

Fig. 108.—Transverse Section of Young Root of *Calamites*

w, Wood of axis; *l*, spaces in the lacunar cortex, whose radiating strands *r* are somewhat crushed; *ex*, outermost cells of cortex with thickened wall.

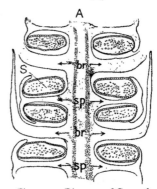

Fig. 109.—Diagram of Cone of *Calamites*

A, Main axis; *br*, sterile bracts; *sp*, sporophylls with four sporangia s attached to each, of which two only are seen.

(see fig. 108). Indeed, so unlike the old roots and the stems are they, that for long they were called by another name and supposed to be submerged stems, but their connection with *Calamites* is now quite certain. As their woody axis develops, the secondary tissue increases and pushes off the lacelike cortex, and the roots become very similar in their anatomy to the stems. Both have similar zones of secondary wood, but the roots do not have those primary canals which are so characteristic of the stems, and thereby they can be readily distinguished from them.

The fructifications of the Calamites were not unlike those of the living types of the family, though in some

PAST HISTORIES OF PLANT FAMILIES 151

respects slightly more complex. Round each cone axis developed rings of sporophylls which alternated with sterile sheathing bracts. Each sporophyll was shaped like a small umbrella with four spokes, and stood at right angles to the axis, bearing a sporangium at each

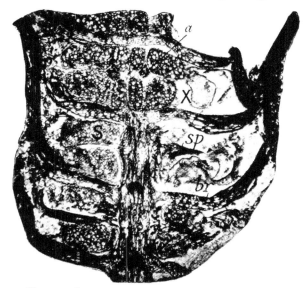

Fig. 110.—Longitudinal Section of Part of *Calamites* Cone

br, Sterile bracts attached to axis; *sp*, attachment of sporophylls; s, sporangia. At X a group of four sporangia is seen round the sporophyll, which is seen at *a*. (Microphoto.)

of the spokes. A diagram of this arrangement is seen in fig. 109.

A photograph of an actual section of such a cone, cut slightly obliquely through the length of the axis, is seen in fig. 110, where the upper groups of sporangia are cut tangentially, and show their grouping round the sporophyll to which they are attached.

A few single tetrads of spores are enlarged in fig. 111, where it will be seen that the large spores are of a similar size, but that the small ones of the tetrads are very irregular. They are aborting members of the tetrad, and appear to have been used as food by the other spores. In each sporangium large numbers of these tetrads develop and all the ripe spores seem to have been of one size.

In a species of *Calamites* (*C. casheana*), otherwise very similar to the common one we have been considering, there is a distinct difference in the sizes of the spores from different sporangia. The small ones, however, were only about one-third of the diameter of the large ones, so that the difference was very much less marked than it was between the small and large spores of the Lycopods.

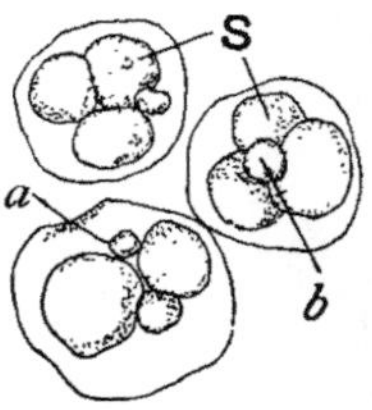

Among the palæozoic members of the group are other genera closely allied to, but differing from *Calamites* in some particulars. One of these is *Archæocalamites*, which has a cone almost identical with that of the living Equisetums, as it has no sterile bracts mingled with the umbrella-like sporophylls. Other genera are more complex than those described for *Calamites*, and even in the simple coned *Archæocalamites* itself the leaves are finely branched and divided instead of being simple scales.

Fig. 111.—Tetrads of Spores of *Calamites*

s, Normal-sized spores; a, b, &c., aborting spores.

But no genus is so completely known as is *Calamites*, which will itself suffice as an illustration of the palæozoic Equisetaceæ. Though the genus, as was pointed out above, shows several important characters differing from those of Equisetum, and parallel to some extent to those of the palæozoic Lycopods, yet these features are more of a physiological nature than a systematic one, and they throw no light on the origin of the family or on its connection with the other Pteridophytes. It is in the extinct family dealt with in the next chapter that we find what some consider as a clue to the solution of these problems.

CHAPTER XVI

PAST HISTORIES OF PLANT FAMILIES

IX. Sphenophyllales

The group to which *Sphenophyllum* belongs is of considerable interest and importance, and is, further, one of those extinct families whose very existence would never have been suspected had it not been discovered by fossil botanists. Not only is the family as a whole extinct, it also shows features in its anatomy which are not to be paralleled among living stems. *Sphenophyllum* became extinct in the Palæozoic period, but its interest is very real and living to-day, and in the peculiar features of its structure we see the first clue that suggests a common ancestor for the still living groups of Lycopods and Equisetaceæ, which now stand so isolated and far apart.

Before, however, we can consider the affinities of the group, we must describe the structure of a typical plant belonging to it. The genus *Sphenophyllum* includes several species (for which there are no common English names, as they are only known to science) whose differences are of less importance than their points of similarity, so that one species only, *S. plurifoliatum*, will be described.

We have a general knowledge of the external appearance of *Sphenophyllum* from the numerous impressions of leaves attached to twigs which are found in the rocks of the Carboniferous period. These impressions present a good deal of variety, but all have rather delicate stems with whorls of leaves attached at regular intervals. The specimens are generally easy to recognize from the shape of the leaves, which are like broad wedges attached at the point (see fig. 112). In some cases the leaves are more finely divided and less fanlike, and it may even happen that on the same branch some may be wedge-shaped like those in fig. 112, and others almost hairlike. This naturally suggests comparison with water plants, which have finely divided submerged leaves and expanded aerial ones. In the case of *Sphenophyllum*, however, the divided leaves sometimes come at the upper ends of the stems, quite near the cones, and so can hardly have been those of a submerged part. The very delicate stems and some points in their internal anatomy suggest that the plant

Fig. 112.—Impression of *Sphenophyllum* Leaves attached to the Stem, showing the wedge-shaped leaflets arranged in whorls

was a trailing creeper which supported itself on the stouter stems of other plants.

The stems were ribbed, but unlike those of the Calamites the ribs ran straight down the stem through the nodes, and did not alternate there, so that the bundles at the node did not branch and fuse as they did in *Calamites*.

The external appearance of the long slender cones was not unlike that of the Calamite cones, though their internal details showed important distinctions.

In one noticeable external feature the plants differed from those of the last two groups considered, and that was in their *size*. Palæozoic Lycopods and Equisetaceæ reached the dimensions of great trees, but hitherto no treelike form of *Sphenophyllum* has been discovered, and in the structure-petrifactions the largest stems we know were less than an inch in diameter.

In the internal anatomy of these stems lies one of the chief interests and peculiarities of the plants. In the very young stage there was a sharply pointed solid triangle of wood in the centre (fig. 113), at each of the corners of which was a group of small cells, the protoxylems. The structure of such a stem is like that of a root, in which the primary wood all grows inwards from the protoxylems towards the centre, and had we had nothing but these isolated young stems it would have been impossible to recognize their true nature.

Such very young stems are rare, for the development of secondary wood began early, and it soon greatly exceeded the primary wood in amount.

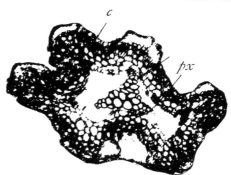

Fig. 113.—*Sphenophyllum*, Transverse Section of Young Stem

c, Cortex, the soft tissue within which has decayed and left a space, in which lies the solid triangle of wood, with the small protoxylem groups *px* at each corner. (Microphoto.)

Fig. 114 shows a photograph of a stem in which the secondary wood is well developed. The primary triangle of wood is still to be seen in the centre, and corresponds to that in fig. 113, while closely fitting to it are the bays of the first-formed secondary wood, which makes the wood mass roughly circular. Outside this the secondary wood forms a regular cylinder round the axis, which shows no sign of annual rings. The cells of the wood are large and approximately square in shape, while at the angles formed at the junction of every four cells is a group of small, thin-walled parenchyma, see fig. 115. There are no medullary rays going

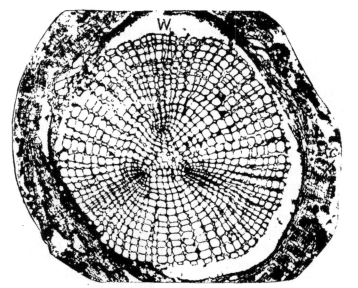

Fig. 114.—*Sphenophyllum*, Transverse Section with Secondary Wood w. At c the cork formation is to be seen. (Microphoto.)

out radially through the wood, such as are found in all other zones of secondary wood, and in this arrangement of soft tissue the plants are unique.

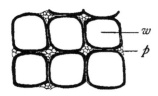

Fig. 115.—Group of Wood Cells w, showing their shape and the small soft-walled cells at the angles between them p

Beyond the wood was a zone of soft tissue and phloem, which is not often preserved, while outside that was the cork, which added to the cortical tissues as the stem grew (see fig. 114, c). Petrified material of leaves and roots is rare, and both are chiefly known through the work of the French palæobotanist Renault. The leaves are chiefly remarkable for the bands of sclerized strengthening tissue, and generally had the structure of aerial, not submerged leaves. The roots were simple in structure,

PAST HISTORIES OF PLANT FAMILIES

and, as in *Calamites*, had secondary tissue like that in the stems.

In the case of the fructifications it is the English material which has yielded the most illuminating specimens. The cones were long and slender, externally covered by the closely packed tips of the scales, which overlapped deeply. Between the whorls of scales lay the sporangia, attached to their upper sides by slender stalks. A diagram will best explain how they were arranged (see fig. 116). Two sporangia were attached to each bract, but their stalks were of different lengths, so that one sporangium lay near the axis and one lay outside it toward the tip of the bract.

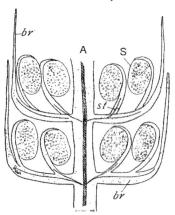

Fig. 116.—Diagram of Arrangement of Scales and Sporangia in Cones of *Sphenophyllum*

A, Axis; *br*, bract; s, sporangium, with stalk *st*.

In its anatomy the stalk of the cone has certain features similar to those in the stem proper, which were among the first indications that led to the discovery that the cone belonged to *Sphenophyllum*. There were numerous spores in each of the sporangia, which had coats ornamented with little spines when they were ripe (fig. 117, if examined with a magnifying glass, will show this). Hitherto the only spores known are of uniform size, and there is no evidence that there was any differentiation into small (male) and large (female) spores such as were found in some of the Lepidodendrons. In this respect *Sphenophyllum* was less specialized than either *Lepidodendron* or *Calamites*.

In the actual sections of *Sphenophyllum* cones the numerous sporangia seem massed together in confusion, but usually some are cut so as to show the attachment of the stalk, as in fig. 117, *st*. As the stalk was long and slender, but a short length of it is usually cut through in any one section, and to realize their mode of attachment to the axis (as shown in fig. 116) it is necessary to study a series of sections.

Of the other plants belonging to the group, *Bowmanites Römeri* is specially interesting. Its sporangia

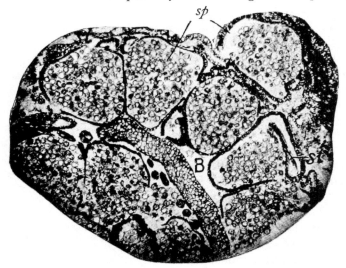

Fig. 117.—Part of Cone of *Sphenophyllum*, showing sporangia *sp*, some of which are cut so as to show a part of their stalks *st*. B, Bract. (Microphoto.)

were borne on stalks similar to those of *Sphenophyllum*, but each stalk had two sporangia attached to it. Two sporangia are also borne on each stalk in *S. fertile*. These plants help in elucidating the nature of the stalked sporangia of *Sphenophyllum*, for they seem to indicate a direct comparison between them and the sporophylls of the Equisetales.

There is, further, another plant, of which we only know the cone, of still greater importance. This cone

PAST HISTORIES OF PLANT FAMILIES 159

(*Cheirostrobus*) is, however, so complex that it would take far too much space to describe it in detail. Even a diagram of its arrangements is extraordinarily elaborate. To the specialist the cone is peculiarly fascinating, for its very complexity gives him great scope for weaving theories about it; but for our purposes most of these are too abstruse.

Its most important features are the following. Round

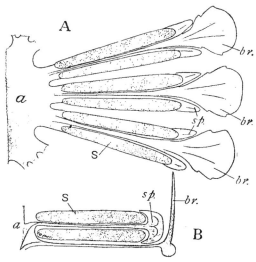

Fig. 118.—A, Diagram of Three-lobed Bract from Cone of *Cheirostrobus*. *a*, Axis; *br*, the three sterile lower lobes of the bract; *sp*, the three upper sporophyll-like lobes, to each of which were attached four sporangia s. B, Part of the above seen in section longitudinal to the axis. (Modified from Scott.)

the axis were series of scales, twelve in each whorl, and each scale was divided into an upper and a lower portion, each of which again divided into three lobes. The lower three of each of these scale groups were sterile and bractlike, comparable, perhaps, with the bracts in fig. 116; while the upper three divisions were stalks round each of which were four sporangia. Each sporophyll segment thus resembled the sporophyll of *Calamites*, while the long sausage-shaped sporangia themselves were more like those of *Lepidodendron*. In fig. 118 is a diagram of a trilobed bract with its three attached sporophylls. Round the axis were very numerous whorls of such bracts, and as the cone was large there were enormous numbers of spore sacs.

A point of interest is the character of the wood of the main axis, which is similar to that of Lepidodendron in many respects, being a ring of centripetally developed wood with twelve projecting external points of protoxylem.

This cone[1] is the most complex fructification of any of the known Pteridophytes, whether living or fossil, which alone ensures it a special importance, though for our purpose the mixed affinities it shows are of greater interest.

To mention some of its characters:—The individual segments of the sporophylls, each bearing four sporangia, are comparable with those of *Calamites*, while the individual sporangia and the length of the sporophyll stalk are similar in appearance to those of *Lepidodendron*. The wood of the main axis also resembles that of a typical *Lepidodendron*. The way the vascular bundles of the bract pass out from the axis, and the way the stalks bearing the sporangia are attached to the sterile part of the bracts, are like the corresponding features in *Sphenophyllum*, and still more like *Bowmanites*.

Many other points of comparison are to be found in these plants, but without going into further detail enough has been indicated to support the conclusion that *Cheirostrobus* is a very important clue to the affinities of the Sphenophyllales and early Pteridophytes. It is indeed considered to have belonged to an ancient stock of plants, from which the Equisetaceæ, and *Sphenophylla*, and possibly also the Lycopods all sprang.

Sphenophyllum, *Bowmanites*, and *Cheirostrobus*, a series of forms that became extinct in the Palæozoic, remote in their structure from any living types, whose

[1] For fuller description of this interesting cone, see Scott's *Studies*, p. 114 *et seq.*

existence would have been entirely unsuspected but for the work of fossil botany, are yet the clues which have led to a partial solution of the mysteries surrounding the present-day Lycopods and Equisetums, and which help to bridge the chasm between these remote and degenerate families.

CHAPTER XVII

PAST HISTORIES OF PLANT FAMILIES

X. The Lower Plants

In the plant world of to-day there are many families including immense numbers of species whose organization is simpler than that of the groups hitherto considered. Taken all together they form, in fact, a very large proportion of the total number of living species, though the bulk of them are of small size, and many are microscopic.

These "lower plants" include all the mosses, and the flat green liverworts, the lichens, the toadstools, and all the innumerable moulds and parasites causing plant diseases, the green weeds growing in water, and all the seaweeds, large and small, in the sea, the minute green cells growing in crevices of the bark of trees, and all the similar ones living by millions in water. Truly a host of forms with an endless variety of structures.

Yet when we turn to the fossil representatives of this formidable multitude, we find but few. Indeed, of the fossil members of all these groups taken together we know less that is of importance and real interest than we do of any single family of those hitherto considered. The reasons for this dearth of fossils of the lower types are not quite apparent, but one which may have some bearing on it is the difficulty of mineralization. It is self-evident that the more delicate and soft-walled any structure is the less chance has it of being preserved without decay long enough to be fossilized. As will have been understood from Chapter II, even when the process of fossilization took place, geologically speaking, rapidly, it can never have been actually accomplished quickly as compared with the counter processes of decay. Hence all the lower plants, with their soft tissue and lack of wood and strengthening cells, seem on the face of it to stand but little chance of petrifaction.

There is much in this argument, but it is not a sufficient explanation of the rarity of lower plant fossils. All through the preceding chapters mention has been made of very delicate cells, such as pith, spores, and, even germinating spores (see fig. 47, p. 68), with their most delicate outgrowing cells. If then such small and delicate elements from the higher plants are preserved, why should not many of the lower plants (some of which are large and sturdy) be found in the rocks?

As regards the first group, the mosses, it is probable that they did not exist in the Palæozoic period, whence our most delicately preserved fossils are derived. There seems much to support the view that they have evolved comparatively recently although they are less highly organized than the ferns. Quite recently experiments have been made with their near allies the liverworts, and those which were placed for one year under conditions similar to those under which plant petrifaction took place, were found to be perfectly preserved at the end of the period; though they would naturally decay rapidly under usual conditions. This shows that Bryophyte cells are not peculiarly incapable of preservation as fossils, and adds weight to the negative evidence of the rocks, strengthening the presumption of their late origin.

That some of the lower plants, among the very lowest and simplest, can be well preserved is shown in the case of the fossil fungi which often occur in microscopic sections of palæozoic leaves, where they infest the higher plants as similar parasitic species do to-day.

PAST HISTORIES OF PLANT FAMILIES 163

We must now bring forward the more important of the facts known about the fossils of the various groups of lower plants.

BRYOPHYTES.—*Mosses.* Of this family there are no specimens of any age which are so preserved as to show their microscopical structure. Of impressions there are a few from various beds which show, with more or less uncertainty in most cases, stems and leaves of what appear to be mosses similar to those now extant, but they nearly all lack the fructifications which would determine them with certainty. These impressions go by the name of *Muscites*, which is a dignified cloak for ignorance in most cases. The few which are quite satisfactory as impressions belong to comparatively recent rocks.

Liverworts are similarly scanty, and there is nothing among them which could throw any light on the living forms or their evolution. The more common are of the same types as the recent ones, and are called *Marchantites*, specimens of which have been found in beds of various ages, chiefly, however, in the more recent periods of the earth's history.

It is of interest to note that among all the delicate tissue which is so well preserved in the "coal balls" and other palæozoic petrifactions, there are no specimens which give evidence of the existence of mosses at that time. It is not unlikely that they may have evolved more recently than the other groups of the "lower" plants.

CHARACEÆ.—Members of this somewhat isolated family (Stoneworts) are better known, as they frequently occur as fossil casts. This is probably due to their character, for even while alive they tend to cover their delicate stems and leaves, and even fruits, with a limy incrustation. This assists fossilization to some degree, and fossil Charas are not uncommon. Usually they are from the recently deposited rocks, and the earliest true Charas date only to the middle of the Mesozoic.

An interesting occurrence is the petrifaction of masses of these plants together, which indicate the existence of an ancient pool in which they must have grown in abundance at one time. A case has been described where masses of *Chara* are petrified where they seem to have been growing, and in their accumulations had gradually filled up the pond till they had accumulated to a height of 8 feet.

The plants, however, have little importance from our present point of view.

FUNGI.—Of the higher fungi, namely, "toadstools", we have no true fossils. Some indications of them have been found in amber, but such specimens are so unsatisfactory that they can hardly afford much interest.

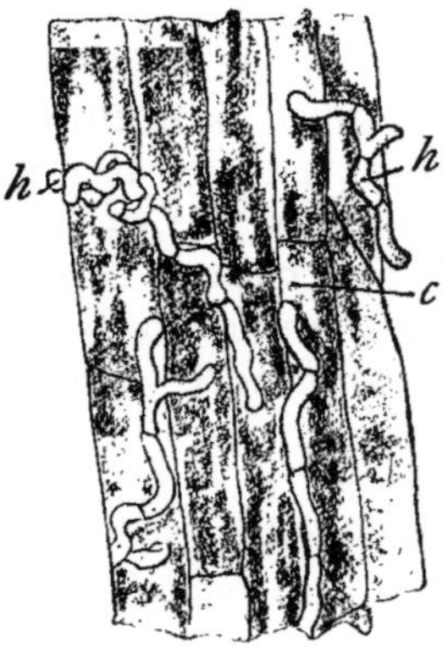

The lower fungi, however, and in particular the microscopic and parasitic forms, occur very frequently, and are found in the Coal Measure fossils. Penetrating the tissues of the higher plants, their hosts, the parasitic cells are often excellently preserved, and we may see their delicate hyphæ wandering from cell to cell as in fig. 119, while sometimes there are attached swollen cells which seem to be sporangia. From the Palæozoic we get leaves with nests of spores of the fungus which had attacked and spotted them as so many do to leaves to-day (see fig. 120). What is specially noticeable about these plants is their similarity to the living forms infesting the higher plants of the present day. Already in the Palæozoic the sharp distinction existed between the

Fig. 119.—The Hyphæ of Fungi Parasitic on a Woody Tree

c, Cells of host; *h*, hyphæ of fungus, with dividing cell walls.

PAST HISTORIES OF PLANT FAMILIES 165

highly organized independent higher plants and their simple parasites. The higher plants have changed profoundly since that time, stimulated by ever-changing surroundings, but the parasites living within them are now much as they were then, just sufficiently highly organized to rob and reproduce.

A form of fungus inhabitant which seems to be useful to the higher plant appears also to have existed in Palæozoic times, viz. *Mycorhiza*. In the roots of many living trees, particularly such as the Beech and its allies, the cells of the outer layers are penetrated by many fungal forms which live in association with the tree and do it some service at the same time as gaining something for themselves. This curious, and as yet incompletely understood physiological relation between the higher plants and the fungi, existed so far back as the Palæozoic period, from which roots have been described whose cells were packed with minute organisms apparently identical with *Mycorhiza*.

Fig. 120.—Fossil Leaf *l* with Nests of Infesting Fungal Spores *f* on its lower side

ALGÆ.—*Green Algæ* (pond weeds). Many impressions have been described as algæ from time to time, numbers of which have since been shown to be a variety of other things, sometimes not plants at all. Other impressions may really be those of algæ, but hitherto they have added practically nothing to our knowledge of the group.

Several genera of algæ coat themselves with calcareous matter while they are alive, much in the same way as do the Charas, and of these, as is natural, there are quite a number of fossil remains from Tertiary and Mesozoic rocks. This is still more the case in the group of the *Red Algæ* (seaweeds), of which the calcareous-coated genera, such as *Corallina* and others, have many fossil representatives. These plants appear so like corals in many cases that they were long held to be of animal nature. The genus *Lithothamnion* now grows attached to rocks, and is thickly encrusted with calcareous matter. A good many species of this genus have been described among fossils, particularly from the Tertiary and Cretaceous rocks. As the plant grew in association with animal corals, it is not always very easy to separate it from them.

BROWN ALGÆ (seaweeds) have often been described as fossils. This is very natural, as so many fossils have been found in marine deposits, and when among them there is anything showing a dark, wavy impression, it is usually described as a seaweed. And possibly it may be one, but such an impression does not lead to much advance in knowledge. From the early Palæozoic rocks of both Europe and America a large fossil plant is known from the partially petrified structure of its stem. There seem to be several species, or at least different varieties of this, known under the generic name *Nematophycus*. Specimens of this genus are found to have several anatomical characters common to the big living seaweeds of the *Laminaria* type, and it is very possible that the fossils represent an early member of that group. In none of these petrified specimens, however, is there any indication of the microscopic structure of reproductive organs, so that the exact nature of the fossils is not determinable. It is probable that though perhaps allied to the Laminarias they belong to an entirely extinct group.

An interesting and even amusing chapter might be written on all the fossils which look like algæ and even have been described as such. The minute river systems that form in the moist mud of a foreshore, if preserved in the rocks (as they often are, with the ripples and raindrops of the past), look extraordinarily like seaweeds—as

PAST HISTORIES OF PLANT FAMILIES

do also countless impressions and trails of animals. In this portion of the study of fossils it is better to have a healthy scepticism than an illuminating imagination.

DIATOMS, with their hard siliceous shells, are naturally well preserved as fossils (see fig. 121), for even if the protoplasm decays the mineral coats remain practically unchanged.

Diatoms to-day exist in great numbers, both in the cold water of the polar regions and in the heat of hot springs. Often, in the latter, one can see them actually being turned into fossils. In the Yellowstone Park they are accumulating in vast numbers over large areas, and in some places have collected to a thickness of 6 feet. At the bottoms of freshwater lakes they may form an almost pure mud of fine texture, while on the floor of deep oceans there is an ooze of diatoms which have been separated from the calcareous shells by their greater powers of resistance to solution by salt water.

Fig. 121.—Diatom showing the Double Siliceous Coat

There are enormous numbers of species now living, and of fossils from the Tertiary and Upper Mesozoic rocks; but, strangely enough, though so numerous and so widely distributed, both now and in these past periods, they have not been found in the earlier rocks.

In one way the diatoms differ from ordinary fossils. In the latter the soft tissues of the plant have been replaced by stone, while in the former the living cell was enclosed in a siliceous case which does not decompose, thus resembling more the fossils of animal shells.

BACTERIA are so very minute that it is impossible to recognize them in ordinary cases. In the matrix of the best-preserved fossils are always minute crystals and granules that may simulate bacterial shapes perfectly. *Bacillus* and *Micrococcus* of various species have been described by French writers, but they do not carry conviction.

As was stated at the beginning of the chapter, from all the fossils of all the lower-plant families we cannot learn much of prime importance for the present purpose. Yet, as the history of plants would be incomplete without mention of the little that is known, the foregoing pages have been added.

CHAPTER XVIII

FOSSIL PLANTS AS RECORDS OF ANCIENT COUNTRIES

The land which to-day appears so firm and unchanging has been under the sea many times, and in many different ways has been united to other land masses to form continents. At each period, doubtless, the solid earth appeared as stable as it is now, while the country was as well characterized, and had its typical scenery, plants, and animals. We know what an important feature of the character of any present country is its flora; and we have no reason to suspect that it was ever less so than it is to-day. Indeed, in the ages before men interfered with forest growth, and built their cities, with their destructive influences, the plants were relatively more important in the world landscape than they are to-day.

As we go back in the periods of geological history we find the plants had an ever-increasing area of distribution. To-day most individual species and many genera are limited to islands or parts of continents, but before the Glacial epoch many were distributed over both America and Europe. In the Mesozoic *Ginkgo* was spread all over the world, and in the present epoch it was confined to China and Japan till it was distributed again by cultivation; while in the Palæozoic period *Lepidodendron* seemed to stretch wellnigh from pole to pole.

The importance of the relation of plant structure to the climate and local physical conditions under which it was growing cannot be too much insisted upon. Modern biology and ecology are continually enlarging and rendering more precise our views of this interrelation, so that we can safely search the details of anatomical structure of the fossil plants for sidelights on the character of the countries they inhabited and their climates.

It has been remarked already that most of the fossils which we have well preserved, whether of plants or animals, were fossilized in rocks which collected under sea water; yet it was also noted that of marine plants we have almost no reliable fossils at all. How comes this seeming contradiction?

The lack of marine plant fossils probably depends on their easily decomposable nature, while the presence of the numerous land plants resulted from their drifting out to sea in streams and rivers, or dropping into the still salt marshes where they grew. Hence, in the rocks deposited in a sea, we have the plants preserved which grew on adjacent lands. In fresh water, also, the plants of the neighbourhood were often fossilized; but actually on the land itself but little was preserved. The winds and rains and decay that are always at work on a land area tend to break down and wash away its surface, not to build it up.

There are many different details which are used in determining the evidence of a fossil plant. Where leaf impressions are preserved which exhibit a close similarity to living species (as often happens in the Tertiary period), it is directly assumed that they lived under conditions like those under which the present plants of that kind are living; while, if the anatomy is well preserved (as in the Palæozoic and several Mesozoic types), we can compare its details with that of similar plants growing under known conditions, and judge of the climate that had nurtured the fossil plant while it grew.

Previous to the present period there was what is so well known as the Glacial epoch. In the earthy deposits of this age in which fossils are found plants are not uncommon. They are of the same kind as those now growing in the cold regions of the Arctic circle, and on the heights of hills whose temperature is much lower than that of the surrounding lowlands. Glacial epochs occurred in other parts of the world at different times; for example, in South Africa, in the Permo-Carboniferous period, during which time the fossils indicate that the warmth-loving plants were driven much farther north than is now the case.

It is largely from the nature of the plant fossils that we know the climate of England at the time preceding the Glacial epoch. Impressions of leaves and stems, and even of fruits, are abundant from the various periods of the Tertiary. Many of them were Angiosperms (see Chap. VIII), and were of the families and even genera which are now living, of which not a few belong to the warm regions of the earth, and are subtropical. It is generally assumed that the fossils related to, or identical with, these plants must therefore have found in Tertiary Northern Europe a much warmer climate than now exists. Not only in Northern Europe, but right up into the Arctic circle, such plants occur in Tertiary rocks, and even if we had not their living representatives with which to compare them, the large size and thin texture of their leaves, their smoothness, and a number of other characteristics would make it certain that the climate was very much milder than it is at present, though the value of some of the evidence has been overestimated.

From the Tertiary we are dependent chiefly on impressions of fossils; anatomical structure would doubtless yield more details, but even as it is we have quite enough evidence to throw much light on the physio-

FOSSIL PLANTS AS RECORDS

graphy of the Tertiary period. The causes for such marked changes of climate must be left for the consideration of geologists and astronomers. Plants are passive, driven before great climatic changes, though they have a considerable influence on rainfall, as has been proved repeatedly in India in recent times.

From the more distant periods it is the plants of the Carboniferous, whose structure we know so well, that teach us most. Although there is still very much to be done before knowledge is as complete as we should wish, there are sufficient facts now discovered to correct several popular illusions concerning the Palæozoic period. The "deep, all-enveloping mists, through which the sun's rays could scarcely penetrate", which have taken the popular imagination, appear to have no foundation in fact. There is nothing in the actual structure of the plants to indicate that the light intensity of the climate in which they grew was any less than it is in a smoke-free atmosphere to-day.

Look at the "shade leaves" of any ordinary tree, such as a Lime or Maple, and compare them with those growing in the sunlight, even on the same tree. They are larger and softer and thinner. To absorb the same amount of energy as the more brilliantly lighted leaves, they must expose a larger surface to the light. Hence if the Coal Measure plants grew in very great shade, to supply their large growth with the necessary sun energy we should expect to find enormous spreading leaves. But what is the fact? No such large leaves are known. *Calamites* and *Lepidodendron*, the commonest and most successful plants of the period, had narrow simple leaves with but a small area of surface. They were, in fact, leaves of the type we now find growing in exposed places. The ferns had large divided leaves, but they were finely lobed and did not expose a large continuous area as a true "shade leaf" does; while the height of their stems indicates that they were growing in partial shade—at least, the shade cast by the small-leaved Calamites and Lepidodendrons which overtopped them.

Indeed there is no indication from geological evidence that so late as Palæozoic times there was any great abnormality of atmosphere, and from the internal evidence of the plants then growing there is everything to indicate a dry or physiologically dry[1] sunny condition.

Of the plant fossils from the Coal Measures we have at least two types. One, those commonly found in nodules *in* the coal itself; and the other, nodules in the rocks above the coal which had drifted from high lands into the sea.

The former are the plants which actually formed the coal itself, and from their internal organization we see that these plants were growing with partly submerged roots in brackish swamps. Their roots are those of water plants (see p. 150, young root of Calamite), but their leaves are those of the "protected" type with narrow surface and various devices for preventing a loss of water by rapid transpiration. If the water they grew in had been fresh they would not have had such leaves, for there would have been no need for them to economize their water, but, as we see in bogs and brackish or salt water to-day (which is physiologically usable in only small quantities by the plant), plants even partly submerged protect their exposed leaves from transpiring largely.

There are details too numerous to mention in connection with these coal-forming plants which go to prove that there were large regions of swampy ground near the sea where they were growing in a bright atmosphere and uniform climate. Extensive areas of coal, and geological evidence of still more extensive deposits, show that in Europe in the Coal Measure period there were vast flats, so near the sea level that they were constantly

---

[1] A brackish swampy land is physiologically dry, as the plants cannot use the water. See Warming's *Oecology of Plants*, English edition, for a detailed account of such conditions. For a simple account see Stopes' *The Study of Plant Life*, p. 170.

being submerged and appearing again as débris drifted and collected over them. Such a land area must have differed greatly from the Europe now existing, in all its features. But the whole continent did not consist of these flats; there were hills and higher ground, largely to the north-east, on which a dry land flora grew, a flora where several of the Pteridosperms and *Cordaites* with its allies were the principal plants. These plants have leaves so organized as to suggest that they grew in a region where the climate was bright and dry.

A fossil flora which has aroused much interest, particularly among geologists, is that known as the Glossopteris flora. This Palæozoic flora has in general characters similar to those of the European Permo-Carboniferous, but it has special features of its own, in particular the genus *Glossopteris* and also the genera *Phyllotheca* and *Schizoneura*.

These genera, with a few others, are characteristic of the Permo-Carboniferous period in the regions in the Southern Hemisphere now known by the names of Australasia, South Africa, and South America, and in India. These regions, at that date, formed what is called by geologists "Gondwanaland". In the rocks below those containing the plants there is evidence of glacial conditions, and it is not impossible that this great difference in climate accounts for the differences which exist between the flora of the Gondwanaland region and the Northern Hemisphere. Unfortunately we have not microscopically preserved specimens of the Glossopteris flora, which could be compared with those of our own Palæozoic.[1]

To describe in detail the series of changes through which the seas and continents have passed belongs to the realm of pure geology. Here it is only necessary to point out how the evidence from the fossil plants may afford much information concerning these continents,

[1] The student interested in this special flora should refer to Arber's British Museum *Catalogue of the Fossil Plants of the Glossopteris Flora.*

and as our knowledge of fossil anatomy and of recent ecology increases, their evidence will become still more weighty. Even now, had we no other sources of information, we could tell from the plants alone where in the past continents were snow and ice, heat and drought, swamps and hilly land. However different in their systematic position or scale of evolutionary development, plants have always had similar minute structure and similar physiological response to the conditions of climate and land surface, so that in their petrified cells are preserved the histories of countries and conditions long past.

CHAPTER XIX

CONCLUSION

In the stupendous pageant of living things which moves through creation, the plants have a place unique and vitally important. Yet so quietly and so slowly do they live and move that we in our hasty motion often forget that they, equally with ourselves, belong to the living and evolving organisms. When we look at the relative structures of plants divided by long intervals of time we can recognize the progress they make; and this is what we do in the study of fossil botany. We can place the salient features of the flora of Palæozoic and Mesozoic eras in a few pages of print, and the contrast becomes surprising. But the actual distance in time between these two types of plants is immense, and must have extended over several million years; indeed to speak of years becomes meaningless, for the duration of the periods must have been so vast that they pass beyond our mental grasp. In these periods we find a contrast in the characters of the plants as striking as that in the characters of the animals. Whole families died out, and new ones arose of more complex and advanced

CONCLUSION

organization. But in height and girth there is little difference between the earliest and the latest trees; there seems a limit to the possible size of plants on this planet, as there is to that of animals, the height of mountains, or the depth of the sea. The "higher plants" are often less massive and less in height than the lower—Man is less in stature than was the Dinosaur—and though by no legitimate stretch of the imagination can we speak of brain in plants, there is an unconscious superiority of adaptation by which the more highly organized plants capture the soil they dominate.

It has been noted in the previous chapters that so far back as the Coal Measure period the vegetative parts of plants were in many respects similar to those of the present, it was in the reproductive organs that the essential differences lay. Naturally, when a race (as all races do) depends for its very existence on the chain of individuals leading from generation to generation, the most important items in the plant structures must be those mechanisms concerned with reproduction. It is here that we see the most fundamental differences between living and fossil plants, between the higher and the lower of those now living, between the forest trees of the present and the forest trees of the past. The wood of the palæozoic Lycopods was in the quality and extent and origin of its secondary growth comparable with that of higher plants still living to-day—yet in the fruiting organs how vast is the contrast! The Lycopods, with simple cones composed of scales in whose huge sporangia were simple single-celled spores; the flowering plants, with male and female sharply contrasted yet growing in the same cone (one can legitimately compare a flower with a cone), surrounded by specially coloured and protective scales, and with the "spore" in the tissue of the young seed so modified and changed that it is only in a technical sense that comparison with the Lycopod spore is possible.

To study the minute details of fossil plants it is

necessary to have an elaborate training in the structure of living ones. In the preceding chapters only the salient features have been considered, so that from them we can only glean a knowledge similar to the picture of a house by a Japanese artist—a thing of few lines.

Even from the facts brought together in these short chapters, however, it cannot fail to be evident how large a field fossil botany covers, and with how many subjects it comes in touch. From the minute details of plant anatomy and evolution pure and simple to the climate of departed continents, and from the determination of the geological age of a piece of rock by means of a blackened fern impression on it to the chemical questions of the preservative properties of sea water, all is a part of the study of "fossil botany".

To bring together the main results of the study in a graphic form is not an easy task, but it is possible to construct a rough diagram giving some indication of the distribution of the chief groups of plants in the main periods of time (see fig. 122).

Such a diagram can only represent the present state of our imperfect knowledge; any day discoveries may extend the line of any group up or down in the series, or may connect the groups together.

It becomes evident that so early as the Palæozoic there are nearly as many types represented as in the present day, and that in fact everything, up to the higher Gymnosperms, was well developed (for it is hard indeed to prove that *Cordaites* is less highly organized than some of the present Gymnosperm types), but flowering plants and also the true cycads are wanting, as well as the intermediate Mesozoic Bennettitales. The peculiar groups of the period were the Pteridosperm series, connecting links between fern and cycad, and the Sphenophyllums, connecting in some measure the Lycopods and Calamites. With them some of the still living groups of ferns, Lycopods, and Equisetaceæ were flourishing, though all the species differed from

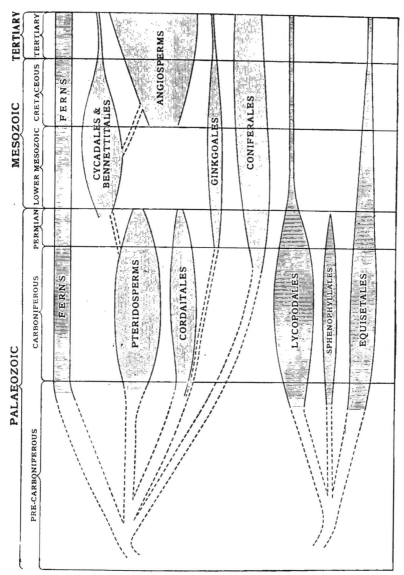

Fig. 122.—Diagram showing the relative distribution of the main groups of plants through the geological eras. The dotted lines connecting the groups and those in the pre-Carboniferous are entirely theoretical, and merely indicate the conclusions reached at present. The size of the surface of each group roughly indicates the part it played in the flora of each period. Those with dotted surface bore seeds, the others spores.

those now extant. This shows us how very far from the beginning our earliest information is, for already in the Palæozoic we have a flora as diversified as that now living, though with more primitive characters.

In Mesozoic times the most striking group is that of the Cycads and Bennettitales, the latter branch suggesting a direct connection between the fern-cycad series and the flowering plants. This view, so recently published and upheld by various eminent botanists, is fast gaining ground. Indeed, so popular has it become among the specialists that there is a danger of overlooking the real difficulties of the case. The morphological leap from the leaves and stems of cycads to those of the flowering plants seems a much more serious matter to presuppose than is at present recognized.

As is indicated in the diagram, the groups do not appear isolated by great unbridged gaps, as they did even twenty years ago. By means of the fossils either direct connections or probable lines of connection are discovered which link up the series of families. At present the greatest gap now lies hedging in the Moss family, and, as was mentioned (p. 163), fossil botany cannot as yet throw much light on that problem owing to the lack of fossil mosses.

This glimpse into the past suggests a prophecy for the future. Evolution having proceeded steadily for such vast periods is not likely to stop at the stage reached by the plants of to-day. What will be the main line of advance of the plants of the future, and how will they differ from those of the present?

We have seen in the past how the differentiation of size in the spores resulted in sex, and in the higher plants in the modifications along widely different lines of the male and female; how the large spore (female) became enclosed in protecting tissues, which finally led up to true seeds (see p. 75), while the male being so temporary had no such elaboration. As the seed

CONCLUSION 179

advances it becomes more and more complex, and when we reach still higher plants further surrounding tissues are pressed into its service and it becomes enclosed in the carpel of the highest flowering plants. After that the seed itself has fewer general duties, and instead of those of the Gymnosperms with large endosperms collecting food before the embryo appears, small ovules suffice, which only develop after fertilization is assured. The various families of flowering plants have gone further, and the whole complex series of bracts and fertile parts which make up a flower is adapted to ensure the crossing of male and female of different individuals. The complex mechanisms which seem adapted for "cross fertilization" are innumerable, and are found in the highest groups of the flowering plants. But some have gone beyond the stage when the individual flowers had each its device, and accomplished its seed-bearing independently of the other flowers on the same branch. These have a combination of many flowers crowded together into one community, in which there is specialization of different flowers for different duties. In such a composite flower, the Daisy for example, some are large petalled and brightly coloured to attract the pollen-carrying insects, some bear the male organs only, and others the female or seed-producing. Here, then, in the most advanced type of flowering plant we get back again to the separation of the sexes in separate flowers; but these flowers are combined in an organized community much more complex than the cones of the Gymnosperms, for example, where the sexes are separate on a lower plane of development.

It seems possible that an important group, if not the dominant group, of flowering plants in the future will be so organized that the individual flowers are very simple, with fewer parts than those of to-day, but that they will be combined in communities of highly specialized individuals in each flower head or cluster.

As well as this, in other species the minute structure

of the vital organs may show a development in a direction contrary to what has hitherto seemed advance. Until recently flowers and their organs have appeared to us to be specialized in the more advanced groups on such lines as encourage "cross fertilization". In "cross fertilization", in fact, has appeared to lie the secret of the strength and advance of the races of plants. But modern cytologists have found that many of the plants long believed to depend on cross fertilization are either self-fertilized or not fertilized at all! They have passed through the period when their complex structures for ensuring cross fertilization were used, and though they retain these external structures they have taken to a simpler method of seed production, and in some cases have even dispensed with fertilization of the egg cell altogether. The female vitality increased, the male becomes superfluous. It is simpler and more direct to breed with only one sex, or to use the pollen of the same individual. Many flowers are doing this which until recently had not been suspected of it. We cannot yet tell whether it will work successfully for centuries to come or is an indication of "race senility".

Whether in the epochs to come flowering plants will continue to hold the dominant position which they now do is an interesting theoretical problem. Flowers were evolved in correlation with insect pollination. One can conceive of a future, when all the earth is under dominion of man, in which fruits will be sterilized for man's use, as the banana is now, and seed formation largely replaced by gardeners' "cuttings".

In those plants which are now living where the complex mechanisms for cross-fertilization have been superseded by simple self-fertilization, the external parts of the more elaborate method are still produced, though they are apparently futile. In the future these vestigial organs will be discarded, or developed in a more rudimentary form (for it is remarkable how organs that were once used by the race reappear in members of it that have

CONCLUSION

long outgrown their use), and the morphology of the flower will be greatly simplified.

Thus we can foresee on both sides much simplified individual flowers—in the one group the reduced individuals associating together in communities the members of which are highly specialized, and in the other the solitary flowers becoming less elaborate and conspicuous, as they no longer need the assistance of insects (the cleistogamic flowers of the Violet, for example, even in the present day bend toward the earth, and lack all the bright attractiveness of ordinary flowers), and perhaps finally developing underground, where the seeds could directly germinate.

In the vegetative organs less change is to be expected, the examples from the past lead us to foresee no great difference in size or general organization of the essential parts, though the internal anatomy has varied, and probably will vary, greatly with the whole evolution of the plant.

But one more point and we must have done. Why do plants evolve at all? Why did they do so through the geological ages of the past, and why should we expect them to do so in the future? The answer to this question must be less assured than it might have been even twenty years ago, when the magnetism of Darwin's discoveries and elucidations seemed to obsess his disciples. "Response to environment" is undoubtedly a potent factor in the course of evolution, but it is not the cause of it. There seems to be something inherent in life, something apparently (though that may be due to our incomplete powers of observation) apart from observable factors of environment which causes slight spontaneous changes, *mutations*, and some individuals of a species will suddenly develop in a new direction in one or other of their parts. If, then, this places them in a superior position as regards their environment or neighbours, it persists, but if not, those individuals die out. The work of a special branch of modern botany seems clearly to indicate the great importance of this seemingly inexplicable spontaneity of life. In environment alone the thoughtful student of the present cannot find incentive enough for the great changes and advances made by organisms in the course of the world's history. The climate and purely physical conditions of the Coal Measure period were probably but little different from those in some parts of the world to-day, but the plants themselves have fundamentally changed. True, their effect upon each other must be taken into account, but this is a less active factor with plants than with men, for we can imagine nothing equivalent to citizenship, society, and education in the plant communities, which are so vital in human development.

It seems to have been proved that plants and animals may, at certain unknown intervals, " mutate "; and mutation is a fine word to express our recent view of one of the essential factors in evolution. But it is a cloak for an ignorance avowedly less mitigated than when we thought to have found a complete explanation of the causes of evolution in " environment ".

In a sketch such as the present, outlines alone are possible, detail cannot be elaborated. If it has suggested enough of atmosphere to show the vastness of the landscape spreading out before our eyes back into the past and on into the future, the task has been accomplished. There are many detailed volumes which follow out one or other special line of enquiry along the highroads and byways of this long traverse in creation. If the bird's-eye view of the country given in this book entices some to foot it yard by yard under the guidance of specialists for each district, it will have done its part. While to those who will make no intimate acquaintance with so far off a land it presents a short account by a traveller, so that they may know something of the main features and a little of the romance of the fossil world.

APPENDIX I

LIST OF REQUIREMENTS FOR A COLLECTING EXPEDITION

In order to obtain the best possible results from an expedition, it is well to go fossil hunting in a party of two, four, or six persons. Large parties tend to split up into detachments, or to waste time in trying to keep together.

Each individual should have strong suitable clothes, with as many pockets arranged in them as possible. The weight of the stones can thus be distributed over the body, and is not felt so much as if they were all carried in a knapsack. Each collector should also provide himself with—

- A satchel or knapsack, preferably of leather or strong canvas, but not of large size, for when the space is limited selection of the specimens is likely to be made carefully.
- One or two hammers. If only one is carried, it should be of a fair size with a square head and strong straight edge.
- One chisel, entirely of metal, and with a strong straight cutting edge.
- Soft paper to wrap up the more delicate fossils, in order to prevent them from scraping each other's surfaces; and one or two small cardboard boxes for very fragile specimens.
- A map of the district (preferably geologically coloured). Localities should be noted in pencil on this, indicating the exact spot of finds. For general work the one-inch survey map suffices, but for detailed work it is necessary to have the six-inch maps of important districts.
- A small notebook. Few notes are needed, but those few *must* be taken on the spot to be reliable.
- A pencil or fountain pen, preferably both.
- A penknife, which, among other things, will be found useful for working out very delicate fossils.

APPENDIX II

TREATMENT OF SPECIMENS

1. The commonest form in which fossils are collected is that which has been described as *impression material* (see p. 12). In many cases these will need no further attention after the block of stone on which they lie has been chipped into shape.

In chipping a block down to the size required it is best to hold it freely in the left hand, protecting the actual specimen with the palm where possible, and taking the surplus edges away by means of short sharp blows from the hammer, striking so that only small pieces come away with each blow. For delicate specimens it is wise to leave a good margin of the matrix round the specimen, and to do the final clearing with a thin-bladed penknife, taking away small flakes of the stone with delicate taps on the handle of the knife.

Specimens from fine sandstones, shales, and limestones are usually thoroughly hard and resistant, and are then much better if left without treatment; by varnishing and polishing them many amateur collectors spoil their specimens, for a coat of shiny varnish often conceals the details of the fossil itself. Impressions of plants on friable shales, on the other hand, or those which have a tendency to peel off as they dry, will require some treatment. In such cases the best substance to use is a dilute solution of size, in which the specimen should soak for a short period while the liquid is warm (not hot), after which it should be slightly drained and the size allowed to dry in. The congealed substance then holds the plant film on to the rock surface and prevents the rock from crumbling away, while it is almost invisible and does not spoil the plant with any excessive glaze.

2. For specimens of *casts* the same treatment generally applies, though they are more apt to separate completely from the matrix after one or two sharp blows, and thus save one the work of picking out the details of their structure.

3. Those blocks which contain *petrifactions*, and can therefore be made to show microscopic details, will require much more treatment. In some cases mere polishing reveals much of the structure—such, for instance, were the "Staarsteine" of the German lapidaries, where the axis and rootlets of a fossil like a treefern show their very characteristic pattern distinctly.

As a rule, however, it is better, and for any detailed work it is essential, to cut thin sections transversely across and longitudinally through the axis of the specimen and to grind them down till they are so transparent that they can be studied through the microscope. The cutting can be done on a lapidary's wheel, where a revolving metal disc set with diamond powder acts as a knife. The comparatively thin slice thus obtained is fastened on to glass by means of hard Canada balsam, and rubbed down with carborundum powder till it is thin enough.

The process, however, is very slow, and an amateur cannot get good results without spending a large amount of time and patience over the work which would be better spent over the study of the plant structures themselves. Therefore it is usually more economical to send specimens to be cut by a professional, if they are good enough to be worth cutting at all, though it is often advisable to cut through an unpromising block to see whether its preservation is such as would justify the expense.

In the case of true "coal balls" much can be seen on the cut surface of a block, particularly if it be washed for a minute in dilute hydrochloric acid and then in water, and then dried thoroughly. The acid acts on the carbonates of which the stone is largely composed, and the treatment accentuates the black-and-white contrast in the petrified tissues (see fig. 10). After lying about for a few months the sharpness of the surface gets rubbed off, as the acid eats it into very delicate irregularities which break and form a smearing powder; but in such a case all that is needed to bring back the original perfection of definition is a quick wash of dilute acid and water. If the specimens are not rubbed at all the surface is practically permanent. Blocks so treated reveal a remarkable amount of detail when examined with a strong hand lens, and form very valuable museum specimens.

The microscope slides should be covered with glass slips (as they would naturally be if purchased), and studied under the microscope as sections of living plants would be.

Microscopic slides of fossils make excellent museum specimens when mounted as transparencies against a window or strong light, when a magnifying glass will reveal all but the last minutiæ of their structure.

4. *Labelling* and numbering of specimens is very important, even if the collection be but a small one. As well as the paper label giving full details, there should be a reference number on every specimen itself. On the microscope slides this can be cut with a diamond pencil, and on the stones sealing wax dissolved

in alcohol painted on with a brush is perhaps the best medium On light-coloured close-textured stones ink is good, and when quite dry can even be washed without blurring.

The importance of marking the stone itself will be brought home to one on going through an old collection where the paper labels have peeled or rubbed off, or their wording been obliterated by age or mould.

A notebook should be kept in which the numbers are entered, with a note of all the items on the paper label, and any additional details of interest.

APPENDIX III

LITERATURE

A short list of a few of the more important papers and books to which a student should refer. The innumerable papers of the specialists will be found cited in these, so that, as they would be read only by advanced students, there is no attempt to catalogue them here.

Carruthers, W., "On Fossil Cycadean Stems from the Secondary Rocks of Britain," published in the *Transactions of the Linnean Society*, vol. xxvi, 1870.

***Geikie, A.**, *A Text-Book of Geology*, vols. i and ii, London, 1903.

Grand'Eury, C., "Flore Carbonifère du département de la Loire et du centre de la France", published in the *Mémoirs de l'Académie des Sciences*, Paris, vol. xxiv, 1877.

***Kidston, R.**, *Catalogue of the Palæozoic Plants in the Department of Geology and Palæontology of the British Museum*, London, 1886.

***Lapworth, C.**, *An Intermediate Text-Book of Geology*, twelfth edition, London, 1888.

Laurent, L., "Les Progrès de la paléobotanique angiospermique dans la dernière decade", *Progressus Rei Botanicæ*, vol. i, Heft 2, pp. 319–68, Jena, 1907.

Lindley, J., and **Hutton, W.**, *The Fossil Flora of Great Britain*, 3 vols., published in London, 1831–7.

Lyell, C., *Principles of Geology* and *The Student's Lyell*, edited by J. W. Judd, London, 1896.

APPENDIX III

Oliver, F. W., and **Scott, D. H.,** "On the Structure of the Palæozoic Seed, *Lagenostoma Lomaxi*", published in the *Transactions of the Royal Society*, series B, vol. cxcvii, London, 1904.

Renault, B., *Cours de Botanique fossile*, Paris, 1882, 4 vols.

Renault, B., *Bassin Houiller et Permien d'Autun et d'Epinac*, Atlas and Text, 1893–6, Paris.

*****Scott, D. H.,** *Studies in Fossil Botany*, London, second edition, 1909.

Scott, D. H., "On the Structure and Affinities of Fossil Plants from the Palæozoic Rocks. On *Cheirostrobus*, a New Type of Fossil Cone from the Lower Carboniferous Strata." Published in the *Philosophical Transactions of the Royal Society*, vol. clxxxix, B, 1897.

*****Seward, A. C.,** *Fossil Plants*, vol. i, Cambridge, 1898.

Seward, A. C., *Catalogue of the Mesozoic Plants in the Department of Geology of the British Museum*, Parts I and II, London, 1894–5.

*****Solms-Laubach, Graf zu,** *Fossil Botany* (translation from the German), Oxford, 1891.

Stopes, M. C., and **Watson, D. M. S.,** "On the Structure and Affinities of the Calcareous Concretions known as 'Coal Balls'", published in the *Philosophical Transactions of the Royal Society*, vol. cc.

*****Stopes, M. C.,** *The Study of Plant Life for Young People*, London, 1906.

*****Watts, W. W.,** *Geology for Beginners*, London, 1905 (second edition).

Wieland, G. R., *American Fossil Cycads*, Carnegie Institute, 1906.

Williamson, W. C., A whole series of publications in the *Philosophical Transactions of the Royal Society* from 1871 to 1891, and three later ones jointly with Dr. Scott; the series entitled "On the Organization of the Fossil Plants of the Coal Measures", Memoir I, II, &c.

Zeiller, R., *Éléments de Paléobotanique*, Paris, 1900.

*****Zittel, K.,** *Handbuch der Palæontologie*, vol. ii; *Palæophytologie*, by Schimper & Schenk, München and Leipzig, 1900.

Those marked * would be found the most useful for one beginning the subject.

GLOSSARY

Some of the more technical terms about which there might be some doubt, as they are not always accompanied by explanations in the text, are here briefly defined.

Anatomy.—The study of the details and relative arrangements of the internal features of plants; in particular, the relations of the different tissue systems.

Bracts.—Organs of the nature of leaves, though not usual foliage leaves. They often surround fructifications, and are generally brown and scaly, though they may be brightly coloured or merely green.

Calcareous.—Containing earthy carbonates, particularly calcium carbonate (chalk).

Cambium.—Narrow living cells, which are constantly dividing and giving rise to new tissues (see fig. 33, p. 57).

Carbonates, as used in this book, refer to the combinations of some earthy mineral, such as calcium or magnesium, combined with carbonic acid gas and oxygen, formula $CaCO_3$, $MgCO_3$, &c.

Carpel.—The closed structure covering the seeds which grow attached to it. The "husk" of a peapod is a carpel.

Cell.—The unit of a plant body. Fundamentally a mass of living protoplasm with its nucleus, surrounded in most cases by a wall. Mature cells show many varieties of shape and organization. See Chapter VI, p. 54.

Centrifugal.—Wood or other tissues developed away from the centre of the stem. See fig. 65, p. 97.

Centripetal.—Wood or other tissues developed towards the centre of the stem. See fig. 65, p. 97.

Chloroplast.—The microscopic coloured masses, usually round, green bodies, in the cells of plants which are actively assimilating.

Coal Balls.—Masses of carbonate of calcium, magnesium, &c., generally of roundish form, which are found embedded in the coal, and contain petrified plant tissues. See p. 28.

GLOSSARY

Concretions.—Roundish mineral masses, formed in concentric layers, like the coats of an onion. See p. 27.

Cotyledons.—The first leaves of an embryo. In many cases packed with food and filling the seed. See fig. 58.

Cross Fertilization.—The fusion of male and female cells from different plants.

Cuticle.—A skin of a special chemical nature which forms on the outer wall of the epidermis cells. See p. 54, fig. 21.

Earth Movements.—The gradual shifting of the level of the land, and the bending and contortions of rocks which result from the slow shrinking of the earth's surface, and give rise to earthquakes and volcanic action.

Embryo.—The very young plant, sometimes consisting of only a few delicate cells, which results from the divisions of the fertilized egg cell. The embryo is an essential part of modern seeds, and often fills the whole seed, as in a bean, where the two fleshy masses filling it are the two first leaves of the embryo. See fig. 58, p. 77.

Endodermis.—The specialized layer of cells forming a sheath round the vascular tissue. See p. 55.

Endosperm.—The many-celled tissue which fills the large "spore" in the Gymnosperm seed, into which the embryo finally grows. See fig. 57.

Epidermis.—Outer layer of cells, which forms a skin, in the multicellular plants. See fig. 21, p. 54.

Fruit.—Essentially consisting of a seed or seeds, enclosed in some surrounding tissues, which may be only those of the carpel, or may also be other parts of the flower fused to it. Thus a peapod is a *fruit*, containing the peas, which are seeds.

Gannister.—A very hard, gritty rock found below some coal seams. See p. 25.

Genus.—A small group within a family which includes all the plants very like each other, to which are all given the same "surname"; e.g. *Pinus montana, Pinus sylvestris, Pinus Pinaster*, &c. &c., are all members of the genus *Pinus*, and would be called "pine trees" in general (see "Species").

Hyphæ.—The delicate elongated cells of Fungi.

Molecule.—The group of chemical elements, in a definite proportion, which is the basis of any compound substance; e.g. two atoms of hydrogen and one atom of oxygen form a molecule of water, H_2O. A lime carbonate molecule (see definition of "Carbonate") is represented as $CaCO_3$.

Monostelic.—A type of stem that contains only one stele.

Morphology.—The study of the features of plants, their shapes and relations, and the theories regarding the origin of the organs.

Nucellus.—The tissue in a Gymnosperm seed in which the large "spore" develops. See figs. 55 and 56, p. 76.

Nucleus.—The more compact mass of protoplasm in the centre of each living cell, which controls its growth and division. See fig. 17, *n.*

Palæobotany.—The study of fossil plants.

Palæontology.—The study of fossil organisms, both plants and animals.

Petiole.—The stalk of a leaf, which attaches it to the stem.

Phloem.—Commonly called "bast". The elongated vessel-like cells which conduct the manufactured food. See p. 57.

Pollen Chamber.—The cavity inside a Gymnosperm seed in which the pollen grains rest for some time before giving out the male cells which fertilize the egg-cell in the seed. See p. 76.

Polystelic.—A type of stem that appears, in any transverse section, to contain several steles. See note on the use of the word on p. 63.

Protoplasm.—The colourless, constantly moving mass of finely granulated, jelly-like substance, which is the essentially living part of both plants and animals.

Rock.—Used by a geologist for all kinds of earth layers. Clay, and even gravel, are "rocks" in a geological sense.

Roof, of a coal seam. The layers of rock—usually shale, limestone, or sandstone—which lie just above the coal. See p. 24.

Sclerenchyma.—Cells with very thick walls, specially modified for strengthening the tissues. See fig. 28, p. 56.

Seed.—Essentially consisting of a young embryo and the tissues round it, which are enclosed in a double coat. See definition of "Fruit".

Shale.—A fine-grained soft rock, formed of dried and pressed mud or silt, which tends to split into thin sheets, on the surface of which fossils are often found.

Species.—Individuals which in all essentials are identical are said to be of the same species. As there are many variations which are not essential, it is sometimes far from easy to draw the boundary between actual species. The specific name comes after that of the genus, e.g. *Pinus montana* is a species of the genus *Pinus*, as is also *Pinus sylvestris*. See "Genus".

Sporangium.—The saclike case which contains the spores. See figs. 52 and 53, p. 75.

GLOSSARY

Spore.—A single cell (generally protected by a cell wall) which has the power of germinating and reproducing the plant of which it is the reproductive body. See p. 75.

Sporophyll.—A leaf or part of a leaf which bears spores or seeds, and which may be much or little modified.

Stele.—A strand of vascular tissue completely enclosed in an endodermis. See p. 62.

Stigma.—A special protuberance of the carpel in flowering plants which catches the pollen grains.

Stomates.—Breathing pores in the epidermis, which form as a space between two curved liplike cells. See fig. 23, p. 54.

Tetrads.—Groups of four cells which develop by the division of a single cell called the "mother cell". Spores and pollen grains are nearly always formed in this way. See p. 75.

Tracheid.—A cell specially modified for conducting or storing of water, often much elongated. The long wood cells of Ferns and Gymnosperms are tracheids.

Underclay.—The fine clay found immediately below some coal seams. See p. 24.

Vascular Tissue.—The elongated cells which are specialized for conduction of water and semifluid foodstuffs.

INDEX

(*Italicized numbers refer to illustrations*)

Abietineæ, 88, 89, 90.
— family characters of, 91.
Algæ, 44, 47, 165.
— brown, 166.
— green, 165.
— red, 165.
Alnus, 85.
Amber, 17.
Amentiferæ, 84.
Anatomy of fossil plants, likeness in detail to that of living plants, 53 et seq.
— — — — differences in detail from that of living plants, 69 et seq.
Andromeda, 84.
Angiosperms, comparison with Bennettitales, 103.
— early history of, 79 et seq.
— general distribution of in time, 177.
— later evolution of, 178.
— male cell of, 52.
Araliaceæ, 85.
Araucareæ, 88, 90, 111.
— description of, 90.
— primitive characters of, 89.
Araucarioxylon, 93, 95.
ARBER, 173.
Archæocalamites, 152.
Artocarpaceæ, 85.
Asterochlaena, 126, 127, *86*, *89*.

Bacillus, 167.
Bacteria, 167.
Baiera, 101.
Bast, 57, *32*.
Bennettitales, 44, 102, 131.
— general distribution of in time, 177.

Bennettites, 103 et seq.
— external appearance of, 103.
— flower-like nature of fructification, 108.
— fructification of, 104, *71*, 105, *72*.
— seed of, 106, *73*.
BERTRAND, 2.
Betula, 85.
Bignonia, 84.
Botryopterideæ, 125, 132.
— description of group, 125.
— fructifications of, 128.
— petioles of, 127, *89*.
— stem anatomy of, 126, *86*, *87*.
— wood of, 128, *90*.
Botryopteris, 126, 127.
— axis with petiole, 127, *88*, *89*.
Bowmanites Römeri, 158, 160.
Brachyphyllum, 89.
BRONGNIART, 2.
Bryophytes, 163.

Calamites, 147, 154, 157, 159, 160, 171.
— branch of, 147, *164*.
— *casheana*, 152.
— cone of, 150, *109*, 151, *110*.
— leaf of, 149, *107*.
— node of, 149, *106*.
— spores of, 152, *111*.
— young roots of, 150, *108*.
— young stem of, 148, *105*.
Cambium, 57, *33*, 65, *43*, 66, 44.
Carbon, film of representing decayed plant, 12.
Carbonate of magnesium, 20
Carbonates of lime, 19.
Carpels, modified leaves, 78.

CARRUTHERS, 186.
Casts of fossil plants, 8, 9, *2*, 10, *3*, *4*, 11, 12.
— of seeds, 11, 12.
— treatment of specimens of, 184.
Casuarina, 83.
Cells, similarity of living and fossil types of, 53.
— principal types of, 53 et seq., *22–33*.
Cell wall, 47, *17*.
Centrifugal wood, 97, *65*.
Centripetal wood, 97, *65*, 116
Chara, 16.
Characeæ, 163.
Cheirostrobus, 159, *118*, 160.
Chloroplast, 47, *17*.
Coal, origin of, 29 et seq.
— of different ages, 33.
— seams in the rocks, 24, *13*, *14*.
— vegetable nature of, 25 et seq.
— 17, Chap. III, p. 22 et seq.
"Coal balls", 18, 19, *10*, 20, 21, 22, 27, *15*, 163, 185.
— — mass of, in coal, 28, *16*.
Coal Measures, climate of, 172.
Companion cells, 57, *32*.
Concretions, 21, 22, 27, *15*, 28.
— concentric banding in, 27, *15*.
Conducting tissue in higher plants, 49, *19*, 50, *20*.
Coniferales, conflicting observations among, 46.
— general distribution in time, 177.
— male cell of, 52.
Corallina, 166.
Cordaiteæ, 39–88, 112.
— comparison of fructifications with those of Taxeæ, 95.
— description of family, 92.
— general distribution in time, 177.
Cordaites, 40, 107, 176.
— fructification of, 95, 96, *64*.
— internal cast of stem, 10, *4*, 93, 94, 95.
— leaves of, once considered to be Monocotyledons, 82, 93.
— leaves of, 93, *61*, 94, *62*A.
— possible common origin with Ginkgo, 102.
— wood of, 94, *62*B.

Cork, 56, *29*.
— cambium, 56, *29*.
Cross fertilization, 179.
Cupresseæ, 88, 90.
— description of, 91.
Cycadeoidea, 103.
Cycads in the Mesozoic period, 40, 41, 42, 113.
— description of group, 109.
— general distribution of in time, 177.
— in Tertiary period, 85.
— large size of male cones of, 110.
— seeds of, 112.
— type of seed of, 76, *57*.
— wood of, 110.
Cycas, 109, 110, 74.
— seed-bearing sporophyll of, 111.
— seeds of, 112, 76.
— comparison with Ginkgo seeds, 112.

DARWIN, 181.
Diatoms, 167, *121*.
Dicotyledons, 41, 44, 79.
— relative antiquity of, 81, 82.
— seed type of, 77, *58*.
Differentiation, commencement of in simple plants, 48.
— of tissues in higher plants, 49, *19*, 50 et seq., *20*.

Embryo of *Ginkgo*, 100.
— in seeds, 76, *57*, 77, *58*.
— of *Bennettites*, 106, *73*.
Endodermis, 55, *26*, 61.
Environment, 181, 182.
Epidermal tissues, 54, *21*, *22*, *23*, 125.
Epidermis cells, fossil impressions of, 13, 14, *8*, 59, *34*, 125.
Equisetales, 44.
— general distribution of in time, 177.
Equisetites, 146, *103*.
Equisetum, 9, 38, 40, 44, 145, 149, 152.
— underground rhizomes of, 43.
Eucalyptus, 83.
Europe, 87, 102.
— ancient climates of, 170.
Evolution, 43.
— in plants, various degrees of in the organs of the same plant, 45 et seq.

INDEX

Evolution in plants, cause of, 181.
— — — suggestions as to possible future lines of, 178.
Expedition, requirements for collecting, 183.
Extinct families, 44

Ferns, sporangia of, 67, *45*.
— connection with Pteridosperms, 123.
— description of group, 124.
— fructifications of among fossils, 131, 132, *92*.
— general distribution of in time, 177.
— germinating spores of, 68, *47*.
Ficus, 83.
Flotsam, 6.
Flowering plant, anatomy of stem, 49, *19*.
Formation of rocks, key to processes, 6.
Fossil plants, indications of ancient climates and conditions, 168.
— — diagram illustration the distribution of, 177, *122*.
Fungi, fossils of, 164.
— parasitic, *119*, 164, 165, *120*.

Gamopetalæ, 84.
Gannister, 25, *14*.
GEIKIE, 186.
Ginkgo, leaf impression, 14, 7, 100, *69*.
— comparison with *Cycas* seeds, 112.
— distribution in the past, 168.
— embryo of, 100.
— epidermis of fossil, 14, *8*, 100.
— foliage of, 99, *66*.
— only living species of genus, 98, *70*.
— possible common origin with *Cordaites*, 102.
— ripe seed of, 99, *67*.
— section of seed of, 100, *68*.
— seed structure of, 76, *57*.
— similarity to *Cordaites*, 96.
Ginkgoaceæ, 88.
Ginkgoales, 88, 98.
— description of group, 98.
— general-distribution in time, 177.
Glacial epoch, 170.
Glossopteris, 173.
Glyptostrobus, 86.
Gondwanaland, 173.

GRAND'EURY, 186.
Gum, 17.
Gymnosperms, 38, 41, 44, 86, 176, 179.
— connection with Pteridosperms, 124.
— general distribution in time, 177.
— relations between the groups of, 88, 89, 90.

Hairs, 54, *22*, 70.
— special forms among fossils, 70.
Heterangium, 119, 122, 123, 127.
— foliage of, 120.
— stem of, 120, *81*.
HOLLICK and JEFFREY, 89.
Horsetails, description of group, 145.
HUTTON, 2.

Impression, form of fossil, *5*, 12, 13, *6*, 14, 7, 15, 80, 81, *59*, 60.
— treatment of specimens of, 184.
Investigators of fossil plants, 2.
Iron sulphide, 20.

Jet, 17.
Juglandaceæ, 83.

Kauri pine, 93.
Kew Gardens, 98.
KIDSTON, 186.

Labelling of specimens, 185.
Lagenostoma, 76, *56*, 118, 119, *80*.
Laminaria, 166.
LAPWORTH, 186.
Latex cells, 55, *27*.
Lauraceæ, 85.
LAURENT, 186.
Leaves, starch manufacture in cells of, 58.
— fossil leaf anatomy, 59, *34*.
— general similarity of living and fossil, 58.
Lepidocarpon, 141, *100*.
Lepidodendron, 9, 10, *3*, 21, *12*, 67, *46*, 72, 75, 134, 144, 145, 157, 160, 171.
— anatomy of stem of, 136, 137, *95*, 138, *96*, 139, 97.
— comparison of reproductive organs with those of living lycopods, 67, *46*.

Lepidodendron, description of, 134.
— distribution in the past, 177.
— fructification of, 139, 140, *98*, 141, *99*.
— huge stumps of, 134, frontispiece.
— leaf bases, 10, *3*, 135, *93*.
— leaf traces of, 139, *97*.
— peculiar fructification of, 75, *54*.
— petrifaction of leaves, 21, *12*.
— rootlike organs of, 69.
— secondary thickening in, 70, *48*, 71, *49*.
— *selaginoides*, stem of, 137, *95*.
— wood of, 70, *48*, 71, *49*.
Liliaceæ, 82.
Limestone, 7, *1*, 24, 25, 36.
LINDLEY, 2, 186.
Literature on fossil plants, 186.
Lithothamnion, 166.
Liverworts, 163.
Lycopods, 38, 40, 42, 44, 67, 133, 175.
— description of group, 133.
— general distribution in time, 177.
— reproductive organs of, 67, *46*.
— secondary wood in fossil, 70, *48*, 71, *49*.
LYELL, 186.
Lyginodendron, 115, 116, 122.
— anatomy of stem of, 116, *78*A.
— petioles of, 117, 118, *79*.
— roots of, 117, *78*B.
— seeds of, 118, 119, *80*.

Magnolia, 83.
Marattia, 130.
Marattiaceæ, 125, 129.
— appearance of, 130.
— description of group, 129.
Marchantites, 163.
Medullosa, 72, 73, 119, 120, 121, *82*, *83*, 122, 123.
— foliage of, 121, 83.
— probable seeds of, 121.
— steles of, 72, *50*, 73, *51*, 120.
Mesozoic, character of flora, 40.
Metaxylem, 57, *31*.
Mycorhiza, 165.
Micrococcus, 167.
Monocotyledons, 41, 44, 79.
— relative antiquity of, 81, 82.
Monostelic anatomy, 63, 126.
Mosses, scarcity of fossils of, 162.

Mosses, fossils of, 163.
Mountain building, from deposits under water, 6.
— — slow and continuous changes, 35.
Muscites, 163.
Mutation, 181.

Nematophycus, 166.
Neuropteris, leaf impression, 6, 13.
— foliage of *Medullosa*, 122.
— with seed attached, 122, *85*.
Nipa, 85.
Nodules, 15, 16, *9*.
Nucleus, 47, *17*.

OLIVER, 187.
Osmunda, 125.
Ovule, word unsuitable for palaeozoic " seeds ", 77.

Palisade cells, 55, *25*.
— tissue in leaves, 58.
— — fossil leaf, 59, *34*.
Palms, 85.
Parenchyma, 55, *24*.
Petrifaction of cells, 4.
Petrifactions, 17.
— of forest débris, 18.
— treatment of specimens of, 184.
Phyllotheca, 173.
Plant, parts of, the same in living and fossil, 59.
— world, main families in, 44.
Platanus, 83.
Polypodiaceæ, 124.
Polystelic anatomy, 63, *72*.
Populus, 83, 85.
Poroxyleæ, 88.
— description of group of, 96.
Poroxylon, anatomy of, 97, 116.
Primitive plants, 46.
Primofilices, 132.
Protococcoideæ, 47, *17*.
Protodammara, 89.
Protoplasm, 47.
Protostele, 62, 70.
Protoxylem, 57, *31*.
Psaronius, 129, 130.
— stem anatomy of, 131, *91*.

INDEX

Pteridophytes, development of secondary wood in fossil forms of, 72.
Pterisosperms, 44, 104, 114, 131.
— description of group, 114 et seq.
— general distribution of in time, 177.
— summary of characters of, 123.
Pteris aurita, 62.

Quarries, 7, *1*.
Quercus, 83, 85.

Race senility, 180.
Ranales, 103.
RENAULT, 2, 156, 187.
Reproductive organs, likeness between those of living and fossil plants, 67, 45, *46*.
— — peculiar characters of some from the Palæozoic, 74.
— — simplicity of essential cells of, 52.
Rocks, persistence of mineral constituents, 36.
— fossils varying in according to the geological age, 37 et seq.
Roof of coal seam, 24, *13*, 25, *14*.
Roots, likeness of structure in living and fossil, 60, *35*.

Salix, 83.
Sambucus, 84.
Schizoneura, 173.
Sclerenchyma, 56, *26*, 59, *34*.
SCOTT, 2, 160, 187.
Secondary wood, development of in fossil members of families now lacking it, 72.
Seeds, series of types from spores to seeds, 75, 76, *52–8*.
— position on the plant, 77, 78.
— Tertiary impressions of, 80, 81, 60.
Selaginella, 75, 133, 134. 7 O
— with four spores in a sporangium, 75, *53*.
Sequoia, 86.
SEWARD, 187.
"Shade leaves", 171.
Shale, 7, *1*, 11, 24, 25, 36.
Sieve tubes, 57, *32*.
Sigillaria, 142, 145.

Sigillaria, cast of leaf bases, 9, *2*, 144, *102*.
— description of, 144.
Silica, 17.
Silicified wood, 17, 80, 87.
SOLMS LAUBACH, 2, 187.
Specimens, treatment of, 184.
Sphenophyllales, 44, 153.
— description of, 153.
— general distribution in time, 177.
Sphenophyllum, 44, 153, 154, 160.
— cone of, 157, 116.
— *fertile*, 158.
— impression of foliage, 154, *112*.
— *plurifoliatum*, 153.
— sporangia of, 158, *117*.
— stem anatomy, 155, *113*, 156, *114*.
— stem in coal ball, 20.
— wood of, 156, *114*, *115*.
Sphenopteris, leaf impression, 11, *5*.
— foliage of Pteridosperms, 115, 77.
Sporangium of ferns, 67, *45*.
— of lycopods, 67, *46*.
— of pteridophytes, 75, *52*, *53*, *54*.
Spores, germinating, in fossil sporangia, 68, *47*.
— peculiar structures among palæozoic examples of, 74.
— series of types from "spores" to "seeds", 75, 76, *52–8*.
— tetrads of, 75, *52*, *53*, *54*.
Sporophyll, 75, *52*, *53*, *54*.
Stangeria, 110.
Stele, modifications of, 62, *36–42*.
Stems, external similarity in living and fossil, 60.
Sternbergia, cast of, 10, *4*.
— pith cast of *Cordaites*, 93.
Stigmaria, 69, 142, 143, 144, 145.
— rootlet of, 143, *101*.
Stomates, 54, *23*.
— in fossil epidermis, 14, *8*.
Stoneworts, 163.
Synclines, 23.

Taxeæ, 88, 90.
— comparison of fructification with that of Cordaiteæ, 95.
— description of, 92.

Taxeæ, fleshy seeds of, 89.
Taxodium, 86.
Taxus, 82.
Time, divisions of geological time, 34.
Tracheides for water storage, 56, 30.
Tree-ferns, 130.
Trigonocarpus, 11, 76, 82, 122, *84*.
— once supposed to be a Monocotyledon, 82.
— probably the seed of Medullosa, 121.
Tubicaulis, 127, *89*.

Unexplored world, 3.
Unicellular plants, 47, *17*.
— — division of cells in, 47, 48, *18*.

"Vascular bundles", relation of to steles, 65, *42*.
— tissue, 57, *31*, *32*, *33*, 59.
— — continued growth of, 65, *43*.
— — importance in plant anatomy, 61 et seq.
Viburnum, 84, 85.

WATTS, 187.
Westphalia, 19.
WIELAND, 2, 102, 187.
WILLIAMSON, 2, 187.
Williamsonia, 104.
Wood, cells composing, 57, *31*.
— centrifugal development of, 97, *65*.
— centripetal development of, 97, *65*.
— parenchyma, 57, *31*.
— silicified, 17, 80, 87.
— solid rings of formed by cambium, 66, *44*.
— vessels of Angiosperms, 58.

Yellowstone Park, 17, 167.
Yew, 82.
Yucca, 82.

ZEILLER, 2, 187.
ZITTEL, 187.
Zygopteris, 127.

By the Same Author

The Study of Plant Life for Young People

Demy 8vo, fully illustrated, 2s. 6d. net

Nature: "Of the various books written for children in elementary schools, *The Study of Plant Life* is quite the most logical and intelligent that we have seen".

Manchester Guardian: "With great skill the knowledge gained from simple experiments in the classroom or laboratory is made effective. . . . Our old friend the cactus . . . and other exotic wonders are happily relegated to very subordinate parts in these bright and attractive pages."

Athenæum: "The author has attained simplicity of language—a more difficult business than writers of textbooks imagine—and as the book is printed in large, clear type, we can strongly recommend it to teachers in schools".

Inquirer: "Is a delightful book, beautifully illustrated".

Scotsman: "The book is a charming guide for the youthful botanist or naturalist to the life history of plants, and it is written with a due regard to the age of those for whom it is intended".

Knowledge: "An excellent book for young people desirous of learning the why and wherefore of plant life".

New Age: "We have seen nothing produced on this side of the Atlantic so good of its kind".

New Phytologist: ". . . may be unhesitatingly pronounced a great success . . . written with a breadth and knowledge he has not met before in an English elementary work".

LONDON: BLACKIE & SON, LIMITED, 50 OLD BAILEY, E.C.

This volume from the
Cornell University Library's
print collections was scanned on an
APT BookScan and converted
to JPEG 2000 format
by Kirtas Technologies, Inc.,
Victor, New York.
Color images scanned as 300 dpi
(uninterpolated), 24 bit image capture
and grayscale/bitonal scanned
at 300 dpi 24 bit color images
and converted to 300 dpi
(uninterpolated), 8 bit image capture.
All titles scanned cover to
cover and pages may include
marks, notations and other
marginalia present in the
original volume.

The original volume was digitized
with the generous support of the
Microsoft Corporation
in cooperation with the
Cornell University Library.

Cover design by Lou Robinson,
Nightwood Design.

Printed in Great Britain
by Amazon